Management Division 2014

Core Programming Area at the 2014 AIChE Annual Meeting

Atlanta, Georgia, USA
16-21 November 2014

ISBN: 978-1-5108-1263-5

Printed from e-media with permission by:

Curran Associates, Inc.
57 Morehouse Lane
Red Hook, NY 12571

CURRAN ASSOCIATES INC. proceedings.com

Some format issues inherent in the e-media version may also appear in this print version.

Printed by Curran Associates, Inc. (2015)

AIChE
120 Wall Street, FL 23
New York, NY 10005-4020

Phone: (800) 242-4363
Fax: (203) 775-5177

www.aiche.org

Additional copies of this publication are available from:

Curran Associates, Inc.
57 Morehouse Lane
Red Hook, NY 12571 USA
Phone: 845-758-0400
Fax: 845-758-2634
Email: curran@proceedings.com
Web: www.proceedings.com

TABLE OF CONTENTS

(66a) Innovation-Working with People and Managing Projects ... 1
Jack Hipple, Eldon R. Larsen

(154a) Strategies for Evaluating Intellectual Property Rights Prior to Product Launch ... 5
Jennifer Roscetti

(154b) Utilization of the Project Management Contractor (PMC) Approach ... 6
Stephen L. Cabano

(154c) The Excitement of Testing and Balancing in the Commissioning Phase of a Project ... 14
Christina M. Borgese

(187a) Introduction to Project Management Fundamentals ... 15
Eldon R. Larsen

(187b) Entrepreneurial Chemical Mega Projects with Parallel Path Process Development ... 16
Marc Privitera

(187d) Becoming a Real Team to Accomplish Projects ... 17
Eldon R. Larsen

Exploiting Polymer-Nanoparticle Interactions to Create a Hydrogel with Biomedical Applications ... 18
Jessica Greer

Validation of High Throughput Electrochemical Gas Sensing Screening System ... 19
Zixuan Wang

Long-Range Correlations in Liquid Water ... 29
Nancy Figueroa

Frictional Study of Polyethylene Glycol Monolayers on Silica Substrate ... 30
Nadiah Nordin

A Machine-Learning Model to Predict Activation Energies of Hydrogenation Reactions ... 31
Jack McCullough

Microscopic Modeling of the Self Assembly of Poly(ethylene oxide)-Poly(propylene oxide)-Poly(ethylene oxide) (PEO-PPO-PEO) Block Copolymers: Critical Micelle Concentrations ... 32
Alexander Colville

(221a) Recommendations for Legislative Actions to Reduce Carbon Emissions in the Electricity Production Sector ... 33
Sam White

(351a) Turning the Big Idea into a Startup ... 61
Dan Watkins

(351c) Starting an Small Engineering Consulting Business – the 10 Year Anniversary of a Startup Company ... 62
Russell F. Dunn

(351d) The Art of the Mass Balance: The Cultural Tools of Entrepreneurial Chemical Startups ... 63
Christina M. Borgese

(408d) The Personal and Family Side of Starting an Independent Business ... 64
Jack Hipple

(408a) Aligning Innovation, Patent Strategies, and Business Models ... 66
Barbara Yuill

(408b) Building an Excellent Entrepreneurial Team ... 67
Marc Privitera

(408c) How to Generate Revenue from Your Technology: Strategies for Licensing Your Intellectual Property ... 68
Jennifer Roscetti

(351f) Designing Winning Boards for Your Startup ... 69
John Finn

(408e) The I-Corps Program at NSF (and NIH) ... 79
Keith McGreggor

Author Index

R&D Management from Beginning to End

Jack Hipple
Principal
Innovation-TRIZ
Tampa, FL

Eldon Larsen
Professor of Engineering
Marshall University
South Charleston, WV

Extended Abstract

This workshop focuses upon managing R&D projects from beginning to end. Beginning refers to mental picture of an idea whereas end refers to solving the customers need and in the process generating profit for the organization. Using a case study, the speakers will delineate challenges involved at every stage of R&D project from a mental image of an idea to a profit making product. The talk also compares old paradigm of innovation management with the new proposed paradigm. The probability of success can be significantly improved using the new paradigm.

Successful Innovation Management

We have seen many idea management flow sheets, gate management processes, and prioritization recommendations starting with an idea that already exists, but no discussion of where the idea came from or what was necessary to generate and frame it. Moreover attention is not paid to shaping and propagation of the idea. This has been especially true over the past decade when processes such as Six Sigma came into vogue whose main thrust was to eliminate deviation. New concepts and innovation are, by their very nature, deviant! In order to generate fresh and innovative ideas we must create a new thought pattern that may challenge the existing status quo in any number of ways. These can even include, at an extreme, replacing one's current products and businesses with totally new approaches and models.

The key elements of successful commercial innovation include:

1. Continuous flow of unmet needs, both from current customers and related to core competencies of the organization ("outside in" as well as "inside out"
2. Using clear mental models of problems and potential solutions
3. Perseverance for idea propagation
4. Clear definition of the business opportunity

5. An appropriate product development and management team
6. Efficient process scale up and continued input from the market
7. Successful project management to insure implementation on schedule and within budget

We will look at each one and discuss the important issues.

1. Continuous Flow of Unmet Needs and Exploration of Uses for Core Competencies

An organization needs to have a process for how it decides what to work on. This may sound simplistic, when this is left to intuition or accident; it is unlikely that the "best" projects will be chosen.

Steps to make sure that an innovation effort is capturing the best opportunities:

- Talk to groups that are not your current customers but that might be able to use the functional capability that you have as a core competence. Customer competitors are another valuable source. The customers of your customers are another excellent source. This insures that you have awareness several levels above your product and customers and alerts you to potential long term needs and trends that may not be obvious to your direct customer or may not be readily disclosed by them.

In these discussions we want to make sure that we have a clear definition of what the "ideal result" that our customers are desiring. We can also look at parallel universes of technology and markets to explore opportunities.

2. Using Clear Mental Models of Problems and Solutions

It is one thing to "invent" a new product, but another to make sure that we understand how a potential customer is going to use the product as well as how the customer's customer is going to use their product. The days are gone when we can just make something, put it in a drum or tank car, and wait for the invoice to be paid. We must be intimately familiar with the product's use, its limitations (including regulatory issues if appropriate). We must understand the details of the physical use of the product and how it affects a customer's products and business.

It is also important to make sure that we do not have blinders on when thinking about use of our product. It is rare that there is only one way to accomplish a function. We must be constantly vigilant about competitive activities, not just from direct competitors, but from parallel approaches to meeting a customer's needs.

3. The People Side

Ideas, innovation, and projects never happen in a vacuum. They are accomplished with teams of people, all of which have different personalities and styles. Recognizing and using these pro-actively is critical to a project's success. How these are measure and used will be discussed.

4. The Business Opportunity

A new product is not a business unless we have a way of sustaining the product's sales, expanding its utility, and developing and enforcing an intellectual property portfolio. As we commercialize a new product, we must be constantly aware of, and be willing to address with resources and commitment, issues such as change in product form, packaging and shipping issues, product purity, and shelf life/stability. These issues are most adequately addressed by constant and continuous communication with the customer.

5. Product Development Teams

We frequently choose product team members based on "who's available". While this may be expedient, it may not serve the best interests of the new business development. The obvious component of team selection is function—marketing R&D, manufacturing, legal, sales, etc. but we frequently ignore the human side of team functioning. If team members spend too much time dealing with personal issues relating to other team members, that energy is not applied to successful new product launch. Tools that can assist in this include measurements such as the Myers Briggs family of instruments, the Hermann Brain Dominance instrument (HBDI), and the Kirton KAI assessment. Knowledge of these personality profiles will not change natural behavior, but can make team members aware of others' preferred styles and open discussion can eliminate hidden barriers to effective team functioning.

6. Efficient process scale up and continued input from the market

Often times, companies strive for launching the best product. It is important to realize that continued input from the market is critical in order to have that 'best product' in the market. It is necessary to efficiently scale up and launch a product sooner than later. It should be perfected using market feedback. Getting market feedback is the only way to close the R&D loop and serve your customers to the fullest.

7. Successful Project Implementation and Management

New product development does not end when the first semi-commercial sample has been OK'd by the customer. A new type of team is now required to oversee the details process design, plant construction (including all the sub issues of permitting, EPA regulations, etc.). Quick and efficient plant start up with appropriate contingency planning is also a part of this effort. Contingency planning, not only related to unforeseen plant construction and start up difficulties, but also to taking care of the customer's needs in these situations must be part of the plan.

Three keys to successful project management lead to successful project completion. These keys involve

- The importance of people
- The importance of defining and setting good objectives
- The importance of good planning

We will apply these three keys to successful project management. This is all about application, not just theory.

386697 Strategies for Evaluating Intellectual Property Rights Prior to Product Launch

Abstract:

Innovation and launch of products involves many parties, including suppliers and finished goods manufacturers. Analyzing the intellectual property ("IP") issues spanning from development to launch is a key component to ensure a successful introduction of a new product. During this session, we will explore what every executive needs to know about evaluating the strength of your company's patents, minimizing risk of infringement, blocking competitors and strategic IP planning. In addition, we will explore recent legislative changes under the America Invents Act ("AIA") that provides new routes for challenging third party patents should a potential risk arise late in the product launch timeline. Specifically, a new Inter Partes Review ("IPR") procedure to attack patents has proved deadly. In the last year almost 95% of patents that entered IPR did not come out the other end. In this session, we will address how to use IPR to your advantage and how to maximize the chances that your own patent survives an IPR attack.

UTILIZATION OF THE PROJECT MANAGEMENT CONTRACTOR (PMC) APPROACH

A
CHEMICAL INDUSTRY DYNAMIC

Stephen L. Cabano
Pathfinder, LLC
7 November 2014

AIChE
The Global Home of Chemical Engineers

Biography: Stephen L. Cabano

- Title: President, Pathfinder, LLC
- Degrees: BS - Mechanical Engineering, Villanova University
- Years of Experience/Professional Field:
  - 25+ years direct project management experience for owner & government clients in the Petroleum, Petrochemical, Chemical, Environmental, Pharmaceutical, Food & Beverage, Mining, Power, Industrial and Commercial industries.
  - As Project Manager/team member in large project teams, has had the responsibility for costs, planning, scheduling, procurement, and similar project-related services.
- Professional affiliation memberships include: Association for the Advancement of Cost Engineering International (AACE), Project Management Institute (PMI), American Institute Chemical Engineering (AIChE), American Society of Training and Development (ASTD), Society of Value Engineers (SAVE); heads Pathfinder's efforts as a product subscriber member of the Construction Industry Institute (CII), and is a 2004-2012 Engineering and Construction Contracting (ECC) Association Board Member, 2006-2007 ECC Association Board Chair.

Presentation Coverage & Objectives

- As in all industry sectors, the Chemical Industry continues to experience organizational and managerial change/evolution
- Presentation will focus briefly on major transitions that have been and are taking place in approaches Chemical Industry Owner organizations use to manage & oversee their major capital projects
  - Review influences driving these transitions
  - Current preferred options & their pros/cons
  - Evolving options

When & Where We Came From

- Chemical Industry Major Capital Project Management & Oversight (MCPMO) up to the 1960s/1970s was generally treated as an in-house/experience driven, customized, company-by-company approach/protocol
- Each Owner considered their approach/protocol to be valuable & proprietary - "intellectual property"

When & Where We Came From

- Downsizing, benchmarking, project complexity & capital project information revolution in mid/late 1970s and beyond led to widespread recognition, documentation & adoption of:
  - Capital Project Performance Ratings
  - Capital Project Development Processes (CPDP)
  - Best Practices and Value Improving Practices
  - Performance Metrics, etc.
  - Increased industry information exchange & awareness
  - Etc.

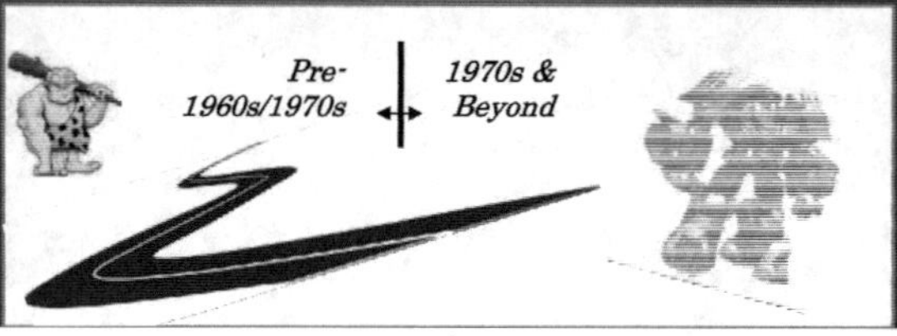

Owner Organization MCPMO CHANGES & Transitions

Option #1 – "Owner Staff" - still most preferred & successful option

- **Prior to mid-1970s –** Owner organizations had extensive internal technical & engineering staffs that managed and provided quality assurance of contracted detailed engineering, procurement & construction Contractor activities
- Used 100% in-house staff of experienced, permanent employee personnel trained in company Policies, Processes, Practices, Procedures, Contracts & Standards

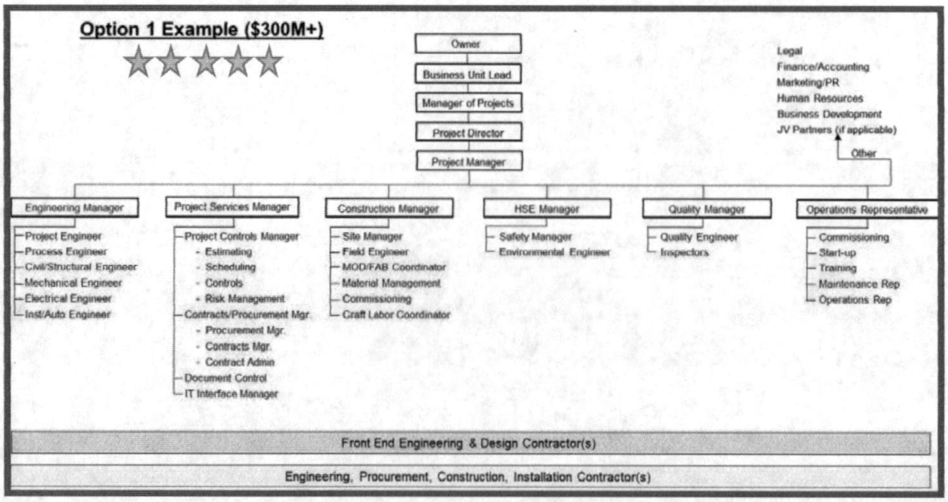

Owner Organization MCPMO CHANGES & Transitions

Option #2 – "Owner Staff + Retirees" – 2nd most preferred & successful option

- **Since mid-1970s** – Owner organizations began downsizing with reduction of internal technical and engineering staffs
- Used mixed teams of Option 1 and experienced former employees (retirees ●) trained in company Policies, Processes, Practices, Procedures, Contracts & Standards

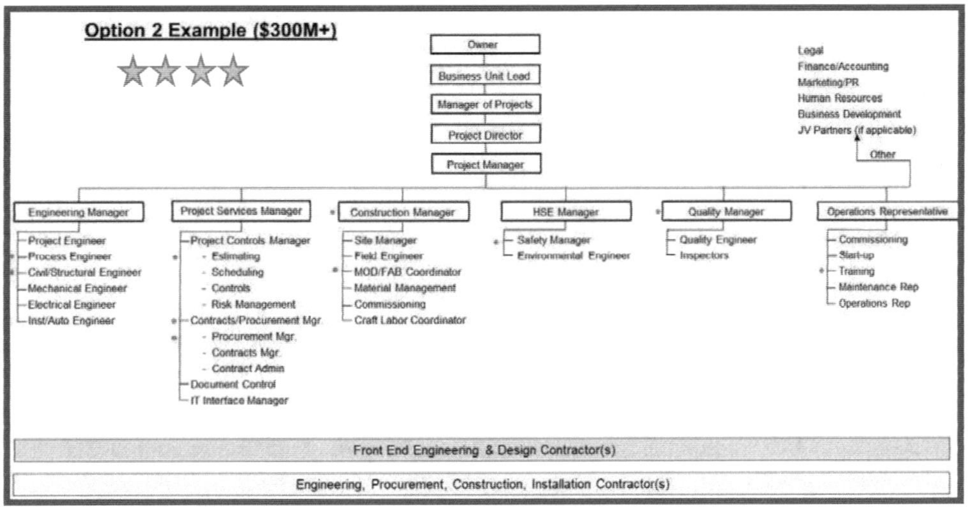

Owner Organization MCPMO CHANGES & Transitions

Option #3 – "Owner + Retiree + Consultant Mix" – 3rd preferred & successful option

- **Since mid-1970s** – Owner organizations continued downsizing with reduction of internal technical & engineering staffs and retirees willing to return diminishes
- Used combination of Options 1 & 2 with pre-screened, qualified, independent expert consultants (●) trained/familiar with company's Policies, Processes, Practices, Procedures, Contracts & Standards

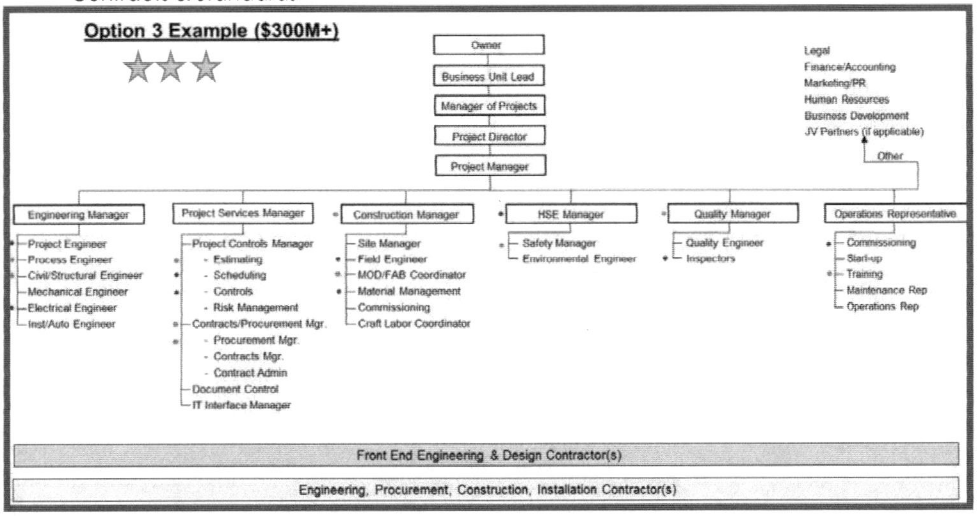

Owner Organization MCPMO CHANGES & Transitions

Option #4 – "Owner Led PMC" – 4th preferred & successful option -- time permitting

- **Since mid-1970s** – Owner organizations continued downsizing with reduction of internal technical & engineering staffs & retirees willing to return diminishes while project number & complexity increase
- Use small, core unit of very senior personnel from Options 1 & 2 to serve as Owner's Management Advisory/Oversight Team (OMAT) to advise and counsel surrogate PMT (•) from experienced Project Management Contractor (PMC) who has recently done work for Owner & is familiar with their Policies, Processes, Practices, Procedures, Contracts & Standards

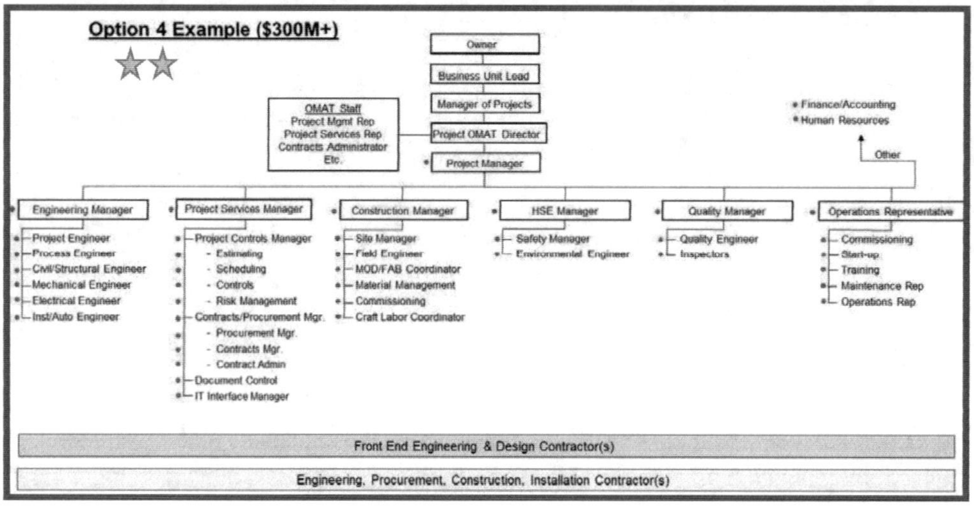

Owner Organization MCPMO CHANGES & Transitions

Option #5 – "Fully Integrated PMC" – 5th & least preferred & successful option -- also requires additional time

- **Since mid-1970s** – Owner organizations continue downsizing with reduction of internal & technical engineering staffs & retirees willingness to return diminishes while project number & complexity increase
- Use fully integrated mix of Options 1 & 2 Owner personnel with an experienced PMC who has recently done work for Owner and is familiar with their Policies, Processes, Practices, Procedures, Contracts & Standards

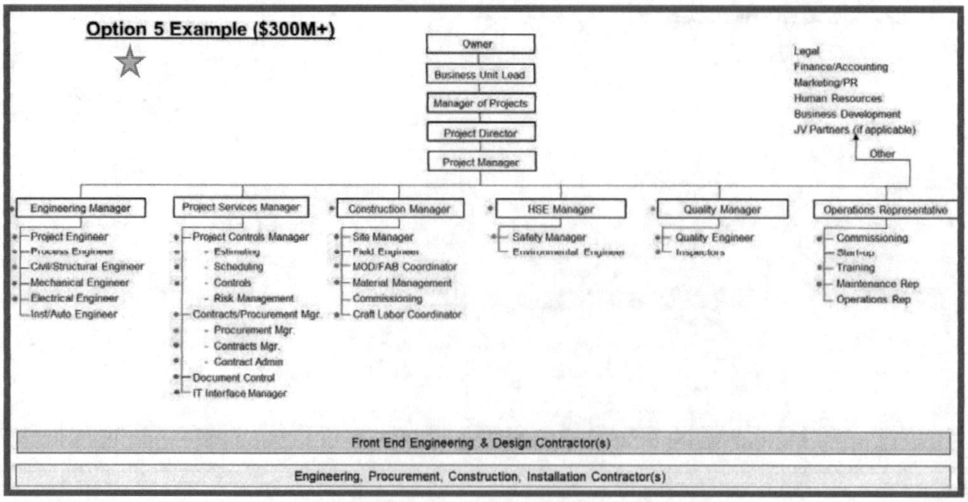

Option Benefits/Disadvantages Comparison

Opt #	PMT Option Title	Benefits	Disadvantages
1	Owner Staff	• Strongest Owner attitude/ commitment • Familiar w/Owner PPPPSC • Best inter-communications • Career enhancement	• Few Owners staffed to implement • Overhead burden during low project workload
2	Owner Staff + Retirees	• Strong Owner attitude/ commitment • Familiar w/Owner PPPPSC • Strong inter-communications	• Diminishing pool of experienced retirees • Willingness to accept long term displacement assignments
3	Owner + Retiree + Consultant Mix	• Consultant expertise enhancement	• Increased 3rd party interfaces • Reduced familiarity w/Owner PPPPSC
4	Owner Led PMC	• Lowest Owner staff demands but with high competencies • If acceptable, can utilize PMC PPPPSC's • PMT homogeneous background except for OMAT	• Increased 3rd party interfaces • Reduced familiarity w/Owner PPPPSC • Owner/Contractor cultural differences • Negative reaction by EPC Contractor(s) • Increased organizational tier
5	Fully Integrated PMC	• Best available personnel from Owner & PMC • Time permitting – selection of optimum PPPPSC	• High conflict potential due to mixed reporting possibilities • Reduced familiarity w/Owner PPPPSC • Owner/Contractor cultural differences • Negative reaction by EPC Contractor(s)

Option Under Development

Evolving Option – "Dedicated PMTC"

- PMTC organizations emerging dedicated solely to Owner PMT support, performing no EPC services and having full portfolio of PPPPSC's for Owner use as desired – Owner oriented experts!
- Allows Owner to customize PMT with some of its own staff if desired, blended appropriately with competent PMTC managerial & functional expert professionals (●) -- reducing execution Contractor competitive concerns as well as Owner's fluctuating project workload overhead burden issues

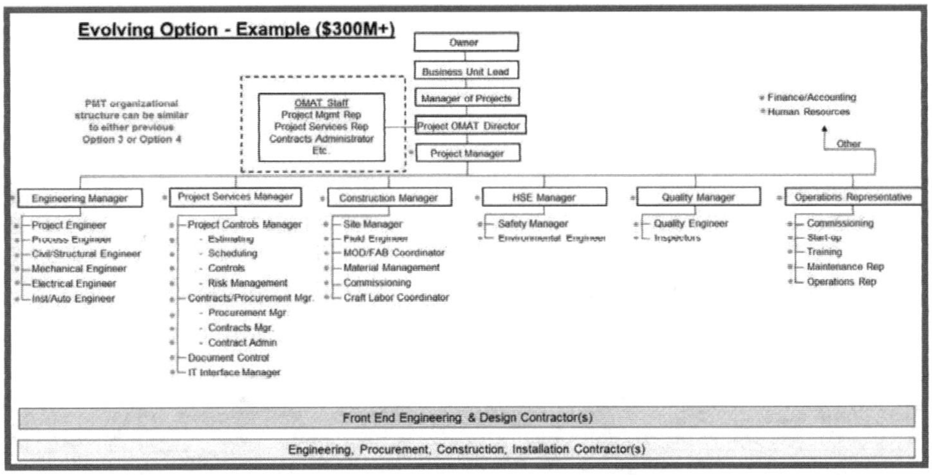

Option Benefits/Disadvantages Comparison

Disadvantages "Evolving Option" may alleviate!!!

Opt #	PMT Option Title	Benefits	Disadvantages
1	Owner Staff	• Strongest Owner attitude/ commitment • Familiar w/Owner PPPPSC • Best inter-communications • Career enhancement	
2	Owner Staff + Retirees	• Strong Owner attitude/ commitment • Familiar w/Owner PPPPSC • Strong inter-communications	
3	Owner + Retiree + Consultant Mix	• Consultant expertise enhancement	• Increased 3rd party interfaces • Reduced familiarity w/Owner PPPPSC
4	Owner Led PMC	• Lowest Owner staff demands but with high competencies • If acceptable, can utilize PMC PPPPSC's • PMT homogeneous background except for OMAT	• Increased 3rd party interfaces • Reduced familiarity w/Owner PPPPSC • Increased organizational tier
5	Fully Integrated PMC	• Best available personnel from Owner & PMC • Time permitting – selection of optimum PPPPSC	• High conflict potential due to mixed reporting possibilities • Reduced familiarity w/Owner PPPPSC

Contact Information

Stephen L. Cabano

Pathfinder, LLC

consulting@pathfinderinc.com

www.pathfinderinc.com

Corporate Office
11 Allison Drive
Cherry Hill, NJ 08003
P: (856) 424 – 7100
F: (856) 424 – 6414

Gulf-Coast Office
16225 Park Ten Place
Suite 500
Houston, TX 77084
P: (281) 292 – 5655
F: (281) 419 – 9977

Calgary *Cherry Hill* *Houston* *Mexico City*

FURTHER REFERENCES

- Industrial Megaprojects: Concepts, Strategies, and Practices for Success
  Edward W. Merrow
  John Wiley & Sons, Inc., 2011
- IPA Newsletter, Q3-(Volume-6,-Issue-3), Sept 2014 – *"Doing More with Less (Without Failing Miserably)"*

387758 The Excitement of Testing and Balancing in the Commissioning Phase of a Project

Abstract:

The most exciting time in a project occurs when approaching mechanical completion as the commissioning begins to ramp up. The Testing and Balancing (TAB) team leaps into action to bridge the overlap between the mechanical completion checklist and the first phases of the commissioning runs. A successful TAB team is an integrated partnership comprised of dedicated and passionate engineering, construction, and operations personnel.

While key members start earlier to shepherd the transition, the multi-organizational TAB team rallies during the Pre Startup Safety Inspection activities. As first fills are loaded and mass starts to move through the system, the action oriented excitement begins as problems are identified, diagnosed, and solved with quick think on your feet agility. Construction personnel coexisting with systems containing chemical loads is a delicate balance requiring planning, training, and communication amongst motivated teammates with clear project leadership skills.

353721 Introduction to Project Management Fundamentals

Abstract:

This presentation introduces the fundamentals of project management. It will be especially useful for new engineers, young professionals, and advanced students, as well as anyone who serves or will served on project teams or leads those teams. We will discuss the importance of people, the importance of setting excellent project objectives, and the importance of good planning. Working with people is perhaps the most difficult. Suggestions will be given on how to better communicate and better achieve unity with project team members. We will discuss the real meaning of the word "Communication" and how that meaning can change the whole approach we take to interacting with people in a productive way. Important ideas and tools for developing excellent project objectives will be presented, including the use of SMART objectives, the triple constraints, and tolerance diagrams. An outline of the project planning process will be reviewed. This planning process includes the use of work breakdown structures, network diagrams, and Gantt charts, as well as resouce histograms. Also, we will briefly discuss doing fatal-flaw analysis.

364220 Entrepreneurial Chemical Mega Projects with Parallel Path Process Development

Abstract:

Entrepreneurial Chemical Mega Projects with Parallel Path Process Development

The start-up world has unleashed the excitement of the American entrepreneurial spirit. The United States leads the world in innovation. The lessons learned that have revolutionized innovation in the manipulation of information can be applied to innovation in chemical mega projects. Manipulating mass, however, is much different than manipulating information. The rules of entrepreneurial efforts have to be applied in a more complex way to insure safety and risk are mitigated within the adventure of changing the world in the chemical industry.

A mega project is a high value, complex, interdependent, multi-organizational undertaking that creates a game changing difference in the landscape of chemical processing. The bonds that are formed, the lessons that are learned and the impact that is felt spans careers, industries and businesses. There is nothing more exciting.

Game changing mega projects are usually based upon some new process development that enables the cost optimization, capacity increase, or yield improvement on some needed chemical product. Feedstock changes, energy efficiency or production routes are usually the chemical engineering technology behind a mega project.

In the old model, process development would be completed, the process work up would be transferred to the process engineering group and the project would move in the serial steps to execution. In today's fast paced market, parallel paths must be considered to accelerate the pace. However, unlike the technology world, the first application out the door, version 1.0, has to be safe and has to work due to the great differences in safety and financial risk between a project that manipulates mass, and a project that manipulates information.

353722 Becoming a Real Team to Accomplish Projects

Abstract:

In project management a major effort goes into becoming a real team! A team is not just a group of individuals! It is a very special group of individuals who have worked and learned to become united in their objectives and willingness to work to accomplish those objectives. This paper will discuss in depth the important principles of people interaction skills needed for people to coalesce into a working unit of synergy. We will discuss communication, what it means, and how to use it. We will talk about leading effective team meetings. We will discuss how to look at issues from other peoples perspectives in order to better understand them and work with them to accomplish objectives. We will discuss listening and barriers to effective listening. All of this will go toward developing a new vision of what it means to be an effective and united team and give important meaning to the word "teamwork."

394223 Exploiting Polymer-Nanoparticle Interactions to Create a Hydrogel with Biomedical Applications

Abstract:

Injectable hydrogels have recently become a relevant material platform in applications such as in vivo drug delivery or as cell carriers because they can be implanted in a minimally invasive manner and closely mimic biological tissue. Attractive properties for such injectable hydrogels are that they easily flow under applied stress (*shear-thinning*) and rapidly heal upon relaxation of the applied stress (*self-healing*). Previously, we developed a shear-thinning and self-healing injectable hydrogel driven by non-covalent interactions between hydrophobically modified cellulose and PEG-PLA nanoparticles. [1] The nature of this gel enables facile loading with a variety of therapeutics as well as tunable release profiles of multiple therapeutics. Specifically, the PEG-PLA nanoparticles allow for loading with hydrophobic therapeutics while a hydrophilic therapeutic can be entrapped within the aqueous phase of the gel.[2] In this work, we have quantified the factors that affect the mechanical properties and release profiles of the gels, demonstrating the range of their tunability. Additionally, we explored other nanoparticles and biopolymers for the fabrication of similar materials. Lastly, we studied the use of this injectable hydrogel as a cell carrier with osteogenic properties. Human mesenchymal stem cells (hMSCs), which are capable of differentiating into bone, cartilage, and fat cells, were encapsulated in the gel and survived injection through a 28 gauge needle into three-dimensional collagen hydrogels. Additionally, PEG-PLA nanoparticles were loaded with dexamethasone to stimulate osteogenic differentiation of the encapsulated hMSCs. We are currently exploring the use of this drug-loaded cell carrier as an injectable system for bone healing. Overall, we present an injectable hydrogel with tunable properties for the controlled delivery of therapeutics with facile synthesis and minimally invasive application *in vivo* that is capable of serving as a stimulatory cell carrier.

---

[1] Appel, E.A., Tibbitt, M.W., Langer, R., Nature Communications., **2014**, *in revision*.

[2] Greer, J.M., Appel, E.A., Tibbitt, M.W., Langer, R., *manuscript in preparation*.

Validation of High Throughput Electrochemical Gas Sensing Screening System

Zixuan Wang

November 17, 2014

UNIVERSITY OF SOUTH CAROLINA

Background

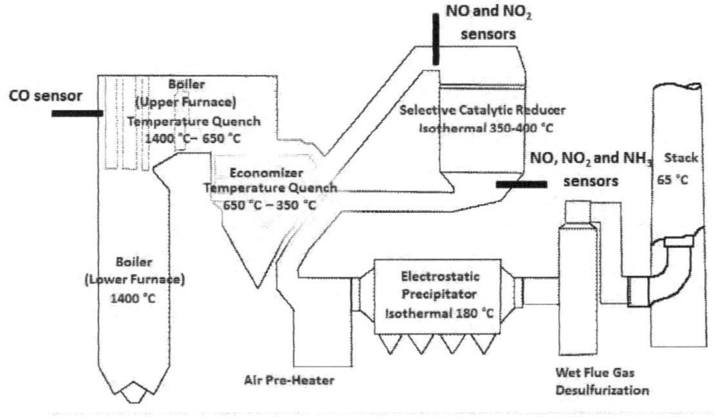

Harsh combustion environment - 800 ° C

Accurate and selective to target gas

UNIVERSITY OF SOUTH CAROLINA

High Throughput Methodology

Subatomic difference affect large scale

Not feasible to individually test material

High throughput synthesis and characterization methods

UNIVERSITY OF SOUTH CAROLINA

High Throughput Methodology

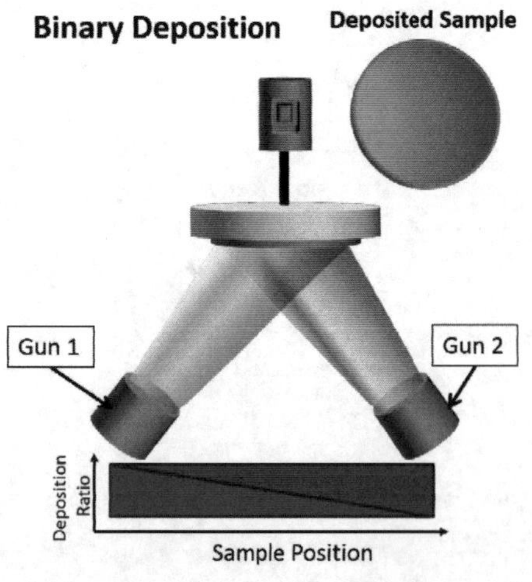

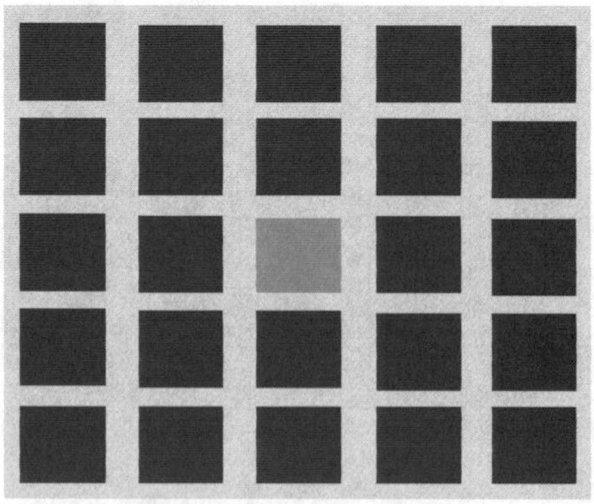

- Normalize affect of distance on the potential

UNIVERSITY OF SOUTH CAROLINA

Development of Uniform Standard Sample

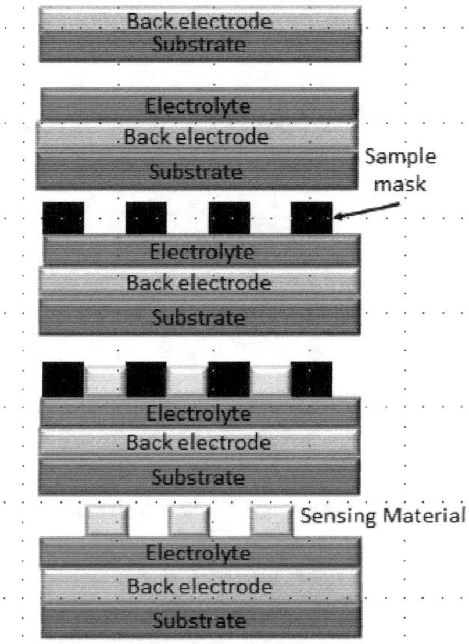

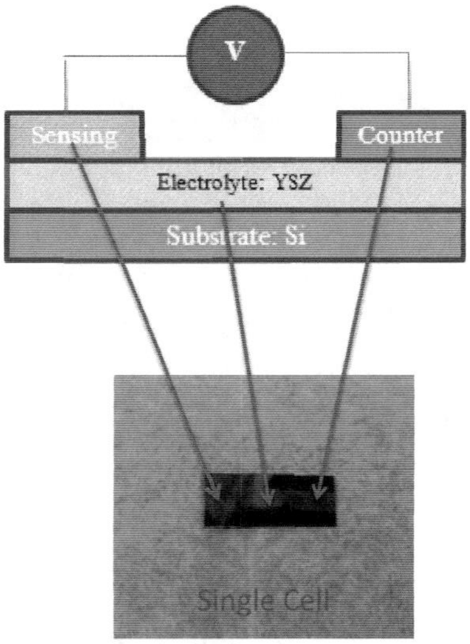

Surface Mechanisms

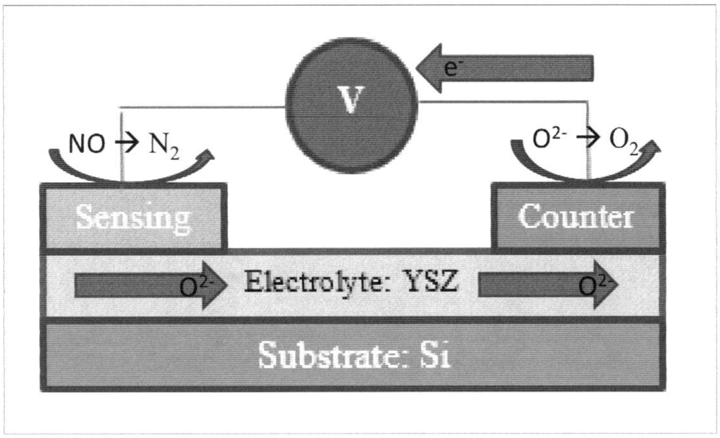

Sensors and Actuators B 121 (2007) 652-663

Reflectometry Results - YSZ

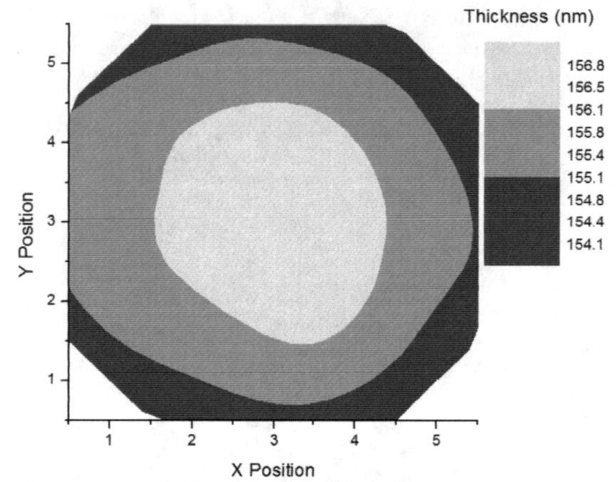

- Results confirm relative uniformity of sample (1.8% difference)

UNIVERSITY OF SOUTH CAROLINA

XRD Results - YSZ

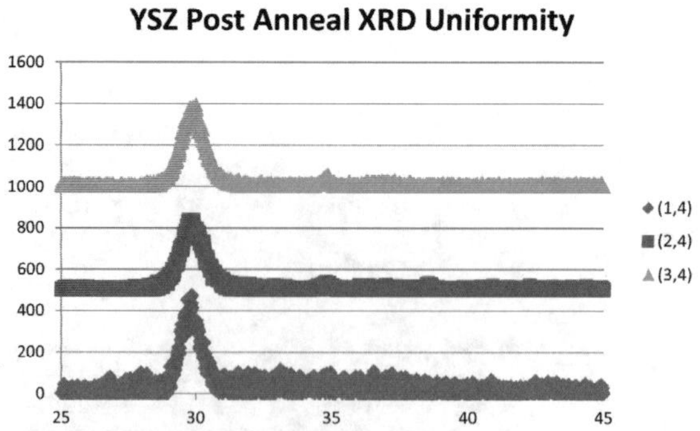

(1,4)	29.745	
(2,4)	29.836	34.848
(3,4)	29.876	34.859

- XRD analysis using Jade
- YSZ Cubic:
  - (111): 29.675°
  - (200): 34.398°
- Confirm identical phases

UNIVERSITY OF SOUTH CAROLINA

Reflectometry Results – WO_3

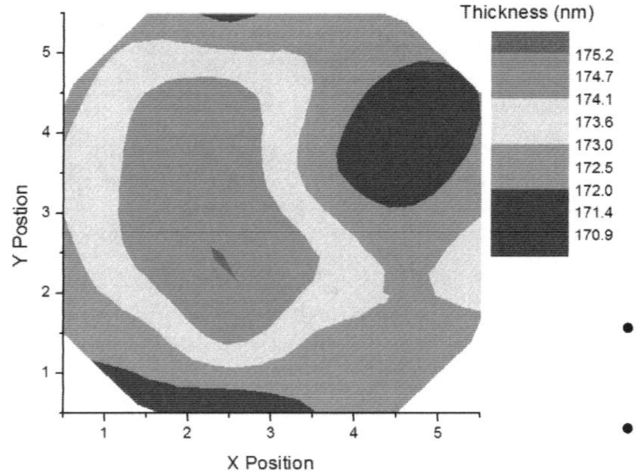

- Confirms relative uniformity (2.5% difference)
- Evidence some parts of the plate that are slightly thicker or thinner

XRD Results – Post Annealed WO_3

(1,4)	23.143	23.653	24.362
(2,4)	23.140	23.671	24.397
(3,4)	23.127	23.641	24.377

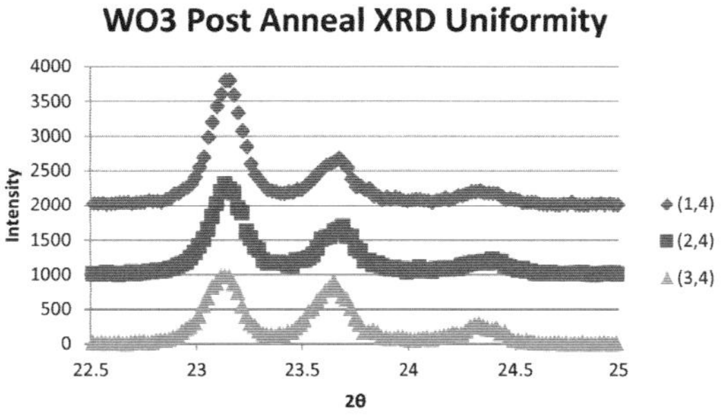

- XRD analysis using Jade
- WO_3 monoclinic:
  - (002): 23.118°
  - (020): 23.582°
  - (200): 24.365°
- Confirm identical phases with phase shifts present

Reflectometry Results - WO_3-YSZ

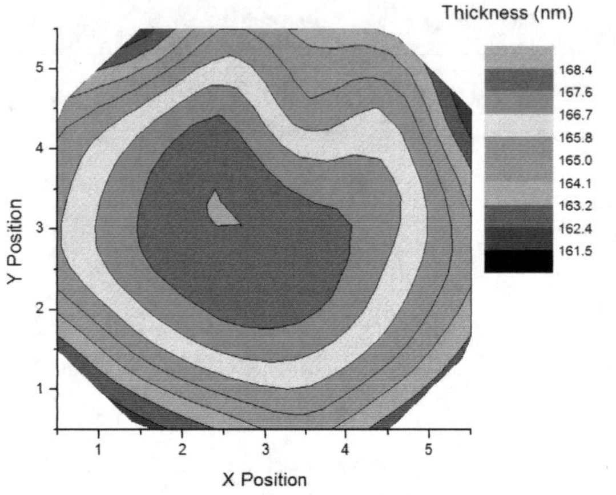

- Sputtering conditions: same
- YSZ held constant
- Confirms relative uniformity (4.3% difference)

XRD Results – WO_3-YSZ Post-annealed

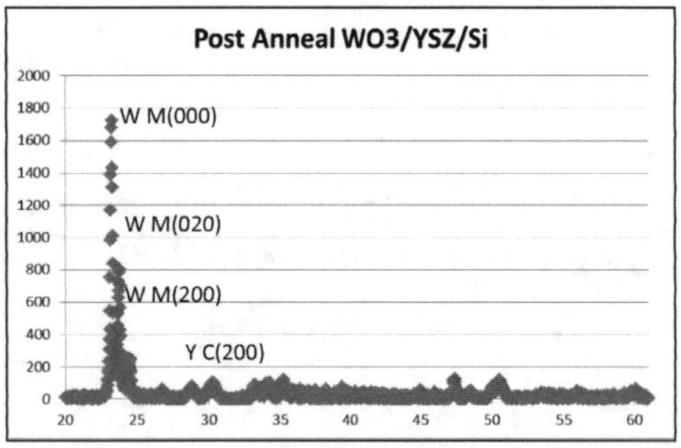

WO_3 monoclinic:

(002): 23.118°

(020): 23.582°

(200): 24.365°

YSZ Cubic:

(111): 29.675°

(200): 34.398°

Electrochemical Cell Setup

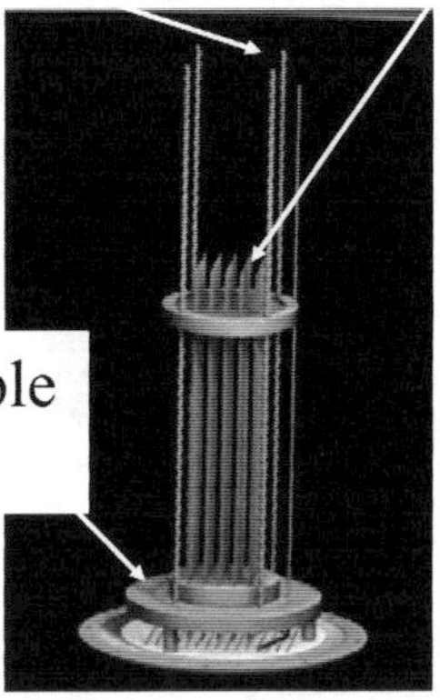

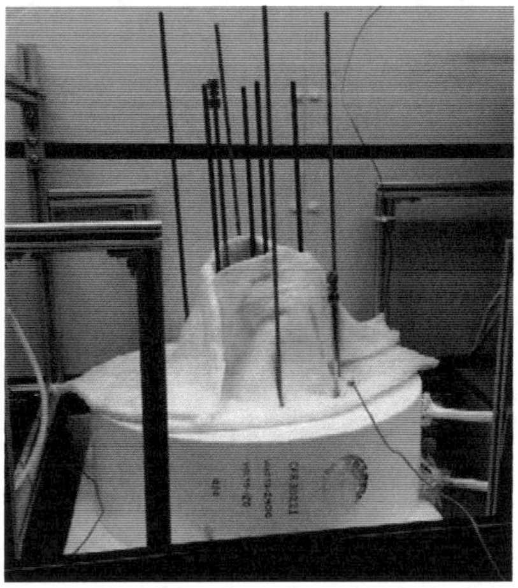

Single Electrochemical Cell Tests

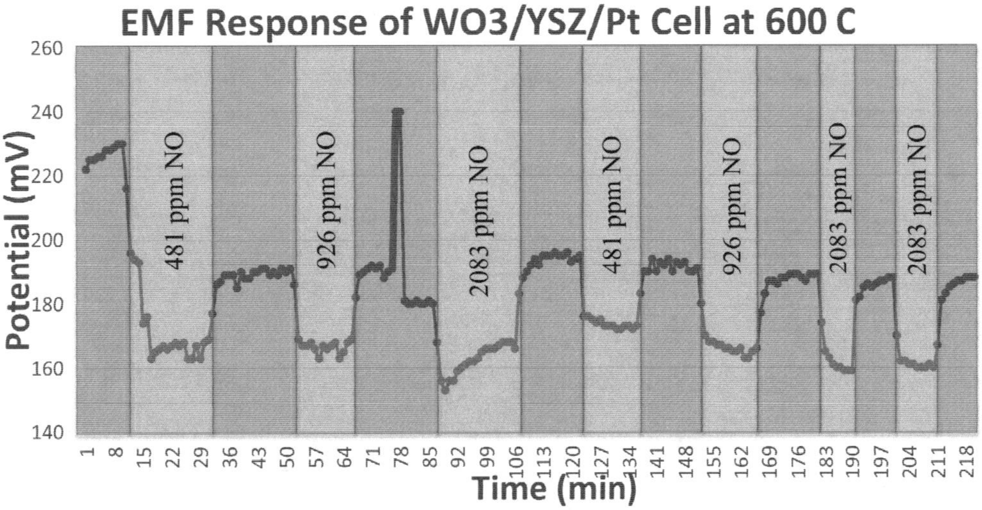

Electrochemical Cell Tests

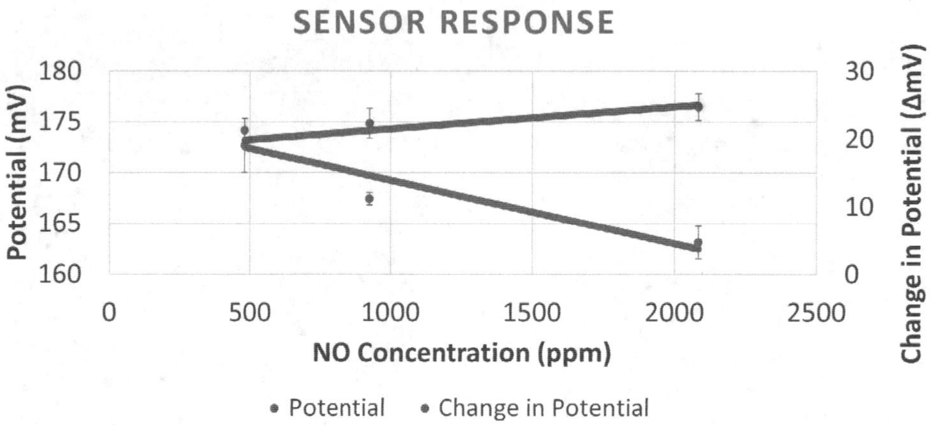

UNIVERSITY OF
SOUTH CAROLINA

Multielectrode Cell Tests

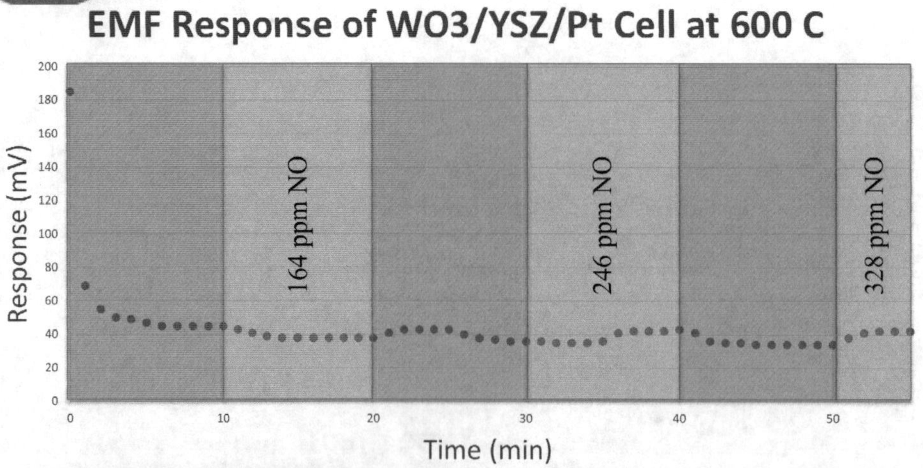

UNIVERSITY OF
SOUTH CAROLINA

Conclusions

- Designed planar sensor array
  - Use for combinatorial screening of electrochemical gas sensors
- Confirm deposition uniformity
  - Reflectometry: thickness
  - XRD: phases
- Cell potential
  - Changes valid use of planar sensor design

UNIVERSITY OF SOUTH CAROLINA

Future Work

- Create multielectrode with increased surface area/perimeter of the electrodes
- Normalize potential to distance of electrodes
- Create multielectrode with different composition gradient

UNIVERSITY OF SOUTH CAROLINA

Acknowledgements

- National Science Foundation (NSF)
- University of South Carolina – College of Engineering and Computing
- Strategic Approaches to the Generation of Electricity (SAGE)
- Principal investigators: Dr. Jochen Lauterbach and Dr. Jason Hattrick-Simpers
- Graduate mentor: Ben Ruiz-Yi

UNIVERSITY OF SOUTH CAROLINA

394216 Long-Range Correlations in Liquid Water

Abstract:

The purpose of this research project was to observe water molecules' environments to determine whether or not the system forms identifiable clusters of similar environments. The different clusters aid in proving the hypothesis that different water patches at differing high and low densities values form in a size of 1nm in a system at room and lower temperatures. Molecular dynamics computer simulations were done using the dl_poly classic program with the water model potential TIP4P/2005. The simulations were run at various configurations for density, temperature, and pressure. The results were read by a program, which took the file from the simulations and determined the coordinate positions of the water molecules. Different clusters were found based on the results. These results are indicative of distinctive structures formed in supercooled water that may lead to the water patches of varying densities.

394155 Frictional Study of Polyethylene Glycol Monolayers on Silica Substrate

Abstract:

Lubricating micro- and nano-electromechanical systems (MEMS and NEMS) has become increasingly challenging as devices are continuously miniaturized. As the dimensions of these systems decrease, their surface area to volume ratios increase significantly and frictional forces become dominant where moving parts exist. Solid-state lubrication via self-assembled monolayers presents a promising alternative to traditional macro-scale lubricants which are too bulky to use in such small systems.[1,2] Previous studies have shown that self-assembled monolayers containing alkylsilanes and mixed alkylsilane and perfluoroalkylsilane chains reduce friction coefficients compared to equal length systems. Although, hydrocarbon-based monolayers have been widely studied and shown to improve frictional properties and device life, mixed monolayers impart the beneficial properties of perfluoroalkylsilanes (low critical surface tension and high thermal and mechanical stability) to the film. Furthermore, a mixed length configuration with longer alkylsilane chains relative to the perfluoroalkylsilane chains leads to a liquid-like layer formed by the taller chains that fold over and align with the shearing direction.[3]

In this study, we investigate polyethylene glycol (PEG) monolayers using molecular dynamics simulations as a potential solid-state lubricant for NEMS. The presence of oxygen atoms within the chains render these chains more flexible than their alkyl- and perfluoroalkyl-counterparts. Thus, even sparsely functionalizing a surface leads to a uniform film. We have investigated the tribological properties of such films across a range of surface coverages and chain lengths. Preliminary results show the effects of surface coverage on friction coefficients is diminished as chain length is increased. This suggests that equivalent friction coefficients could be achieved while reducing separation distances by using longer chains on sparser surfaces.

394001 A Machine-Learning Model to Predict Activation Energies of Hydrogenation Reactions

Abstract:

The goal of this research is to develop a machine-learning model to predict activation energies associated to the hydrogenation of organic compounds in the presence of specific catalytic substrates. The current method of predicting activation energies involves a linear relationship with the change in Gibbs free energy of the reaction, which can be very inaccurate. Other methods include experimentation and simulation, both of which have significant cost and/or time drawbacks. This research seeks to improve on these methods to develop a means to predict activation energies quickly, accurately, and with minimal cost.

393764 Microscopic Modeling of the Self Assembly of Poly(ethylene oxide)-Poly(propylene oxide)- Poly(ethylene oxide) (PEO-PPO-PEO) Block Copolymers: Critical Micelle Concentrations

Abstract:

Single chain mean field theory (SCMFT) is applied as a coarse-grain simulation model for Pluronics, a trademarked branch of triblock copolymers. The microscopic model is used to quantify the micellization process of the nonionic class of surfactants at 37 degrees Celsius for drug delivery applications. The simple model with only three interaction parameters is shown to be able to reproduce with high accuracy experimental data for the critical micelle concentrations of different chain lengths of the polymers. These parameters are physically significant and are close to the experimental values from Flory models. In addition, aggregation number of the surfactants and volume fraction profiles are explored. The effects of the chain lengths of hydrophobic tails and hydrophilic heads on CMC values are discussed. SCMFT is transferable in surfactants of different ABA block conformations and other more complex polymer structures where previous simulations have not excelled.

Recommendations for Legislative Actions to Reduce Carbon Emissions in the Electricity Production Sector

Sam White

American Institute of Chemical Engineers
University of Michigan
Washington Internships for Students in Engineering

August 2014

Executive Summary

There is currently a movement in America pushing for environmental effects to be taken into consideration when assessing new and existing business ventures. As emphasized by President Obama"s 2014 State of the Union Address, a major area for this movement to make progress is within the energy sector [1]. Electricity production currently produces the most greenhouse gas emissions of any energy sector in the US, making it the energy sector where the largest environmental impact can be made with proper policy action [2].

Luckily, currently-available, low-emission renewable energy technologies are most well suited for electricity production over any other energy sector application. As a result of recent government regulations combined with high coal prices, coal plants, which currently dominate the electricity production landscape, are being retired at an accelerated rate [8]. This, combined with increased electricity demand, has opened up the door for new electricity production technology to be implemented. However, due to the low price of natural gas resulting mainly from the onset of fracking, efforts to implement new renewable-sourced electricity production facilities have been stalled in favor of natural gas power plants. While natural gas plants produce only half the carbon emissions of coal, and do not produce harmful byproducts such as sulfur dioxide or mercury, they are not as environmentally friendly as the other available alternatives, such as nuclear, solar, or wind [11] [12] [13] [14].

The fact that natural gas dominates the shift away from coal is not surprising, however, given the current economic landscape. The figure below shows the levelized cost of electricity (LCOE) for select available electricity production technologies, giving a $/Mwh cost of production by factoring in all applicable costs (construction, fuel, maintenance, operation, etc.) [23].

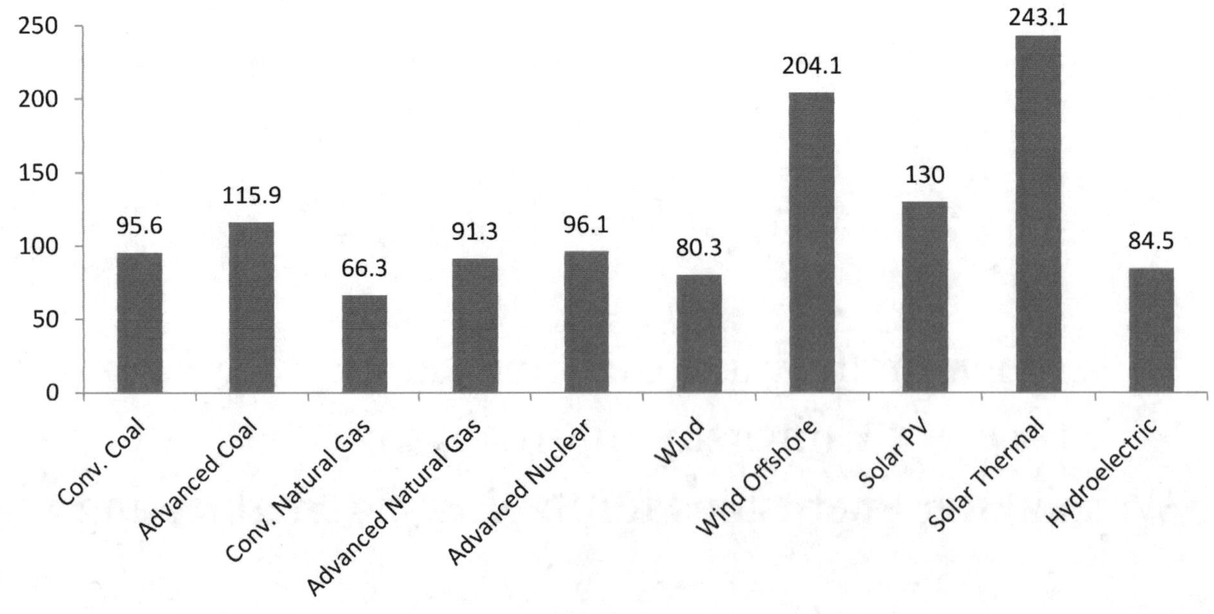

US Average LCOE for Plants Entering Service in 2019, in 2012 $/Mwh [23 EIA LCOE]

As seen in figure above, conventional natural gas currently offers the cheapest form of electricity production. It is important to note, though, that some alternatives like nuclear or wind are already competitive with other conventional electricity production methods like coal. Other alternatives, like solar or offshore wind, are still too expensive to be competitive at the moment, mainly due to the fact that electricity production from renewable sources is still too variable. With the advent of advanced electricity storage technology this issue would be eliminated, but such technology is currently unavailable or too expensive to be implemented economically [19].

Because of their positive environmental impacts, alternative electricity production methods are given some assistance by the federal government in order to make them more economically competitive. These economic aids mainly come in two forms: tax credits and loan programs. On the tax side, an installment tax credit allows solar production facilities a credit equal to 30% of the cost of implementation, while a production tax credit allows wind, hydro, and geothermal plants to receive a .23 cents/kwh credit. However, the installment tax credit is only available until the end of 2016, and the production tax credit already expired at the beginning of 2014 [26]. On the loan program side, the government has extended loans to help both nuclear and renewable power plants, though the renewable plant loan program expired in 2011 [27] [28].

On top of aid being given to alternative electricity production methods, the federal government can also level the playing field by putting a burden on conventional electricity sources. This mainly comes in one of two forms, a carbon tax or a cap-and-trade program, both of which seek to charge electricity producers for the amount of carbon emissions they produce. On top of producing revenue for the federal government, these programs help incentivize electricity producers to innovate new ways to reduce their carbon emissions. Though no such program has ever been implemented, a cap-and-trade program was able to at least pass the House in 2009 [24] [25].

Even small actions by the federal government can have noticeable effects in shifting the production methods used in the electricity production sector. Just a relatively small tax such as $10/metric ton of carbon emissions would cause significant carbon emission reductions compared a scenario in which no government actions are taken [33]. Additional analysis specific to natural gas and renewables shows that any carbon tax would actually increase the amount of natural gas-based electricity production for the next couple of decades, and increase the amount of renewable-based electricity production indefinitely. Additionally, if current favorable tax policies for renewables are renewed long-term, the use of renewables is significantly higher than those projected given current tax policy deadlines, especially after 2030 [34] [35].

As a result, it is recommended that a two-pronged approach be taken to limit the amount of economic disturbance while ensuring the electricity production sector makes a greater shift towards renewables given the reduction of coal use. First, a carbon tax equivalent of $10/metric ton of carbon emissions should be enacted to reflect the social cost of emissions. Each year this tax would increase 5% per year after inflation in order to push companies to innovate new carbon reduction technologies. Second, the production tax credit should be extended for all renewable electricity production methods using the current credit of .23 cents/kwh. This credit would decrease 5% per year after inflation to mirror the carbon tax in order to account for the increased rate of improvement in renewable technologies and the relative advantage given by the increasing carbon tax.

It is important that such a program be long-term in order to allow industry to accurately model the economic landscape their projects will enter. Given the current gridlock in Washington and the long-term nature of implementing new electricity production facilities, a long-term tax plan is the only way to ensure economic stability. Factoring in only revenue from taxes on natural gas use and costs from credits extended to renewables use, such a program would pull in an average surplus of $5.5 billion per year between 2015 and 2040; factoring in coal would increase this figure significantly, particularly in the near future. Feeding at least part of this surplus into loan programs for nuclear plants would ensure that an increased amount of low carbon electricity production facilities are able to come to market.

Table of Contents

Executive Summary

Table of Contents

Table of Figures

Table of Tables

1.0 Introduction

2.0 Overview of Electricity Production Methods and Their Environmental Impacts

- 2.1 Coal and Natural Gas
- 2.2 Nuclear
- 2.3 Solar and Wind
- 2.4 Other Renewables
- 2.5 Environmental Impact Summary Table

3.0 Economics in the Electricity Production Sector

- 3.1 Comparing Costs of Electricity Production Technologies
- 3.2 Capacity Factor

4.0 Recent Policy Strategies

- 4.1 Tax Credits for the Electricity Production Sector
- 4.2 Loan Programs
- 4.3 Carbon Tax and Cap-and-Trade Program

5.0 Projected Electricity Sector Growth: Electricity Sector Capacity Additions for 2013

- 5.1 Policy Action Effects: Overall Carbon Emissions
- 5.2 Policy Action Effects: Conventional Fuels
- 5.3 Policy Action Effects: Renewables

6.0 Policy Recommendations

7.0 References

Table of Figures

Figure 1. Total Greenhouse Gas Emissions by Economic Sector in 2012 [2]

Figure 2. Share of Total US Electricity Production by Source [Adapted from 8]

Figure 3. US Average LCOE for Plants Enetering Service in 2019, in 2012$/MWh [23]

Figure 4. Capacity Factor (%) of Various Electricity Production Technologies [23]

Figure 5. Electricity Generation Capacity Addition Projections by Fuel Type, 2008 (Gw) [30]

Figure 6. Electricity Generation Capacity Addition by Fuel Type 2013 (GW) [Adapted from 31]

Figure 7. Electricity Generation Addition Projections by Fuel Type 2014 (GW) [32]

Figure 8. Electricity-Related CO_2 Emissions in Five Cases, 2000-40 [33]

Figure 9. Natural Gas Electricity Generation in Five Cases, 2000-40 [34]

Figure 10. Renewable Electricity Generation in Four Cases, 2000-40 (billion kWh) [Adapted from 35].

Table of Tables

Table 1. Life-Cycle Environmental Impact Summary of Electricity Production Technologies [Adapted from 11]

1.0 Introduction

In his 2014 State of the Union address, President Obama stated that "[O]ne of the biggest factors in bringing more jobs back is our commitment to American energy"[1]. He stressed the importance of "a cleaner energy economy" and acknowledged that "[it] won't happen overnight, and it will require tough choices along the way." Though he touted the success of his "all-of-the-above energy strategy", and highlighted the advantages of natural gas and solar, the President laid out no definitive plans for how the US would approach energy production in the future. However, with the release of recent carbon emission regulations, the executive branch seems to be taking at least some action to direct the country towards a cleaner energy economy. However, if a cleaner energy economy is going to happen in America, the legislative branch needs to get more involved by developing and implementing a long-term energy policy plan.

Of all the areas of energy production in the America, the most promising area for change is the electricity production sector, since a vast majority of alternative energy sources are most viable for electricity production instead of applications like transportation or residential heating. Luckily, this is also the energy sector in which the most impactful changes can be made: in 2012, electricity production was the largest single contributor to greenhouse gas emissions in the US, accounting for 32% of greenhouse gas emissions as seen in Figure 1 below [2].

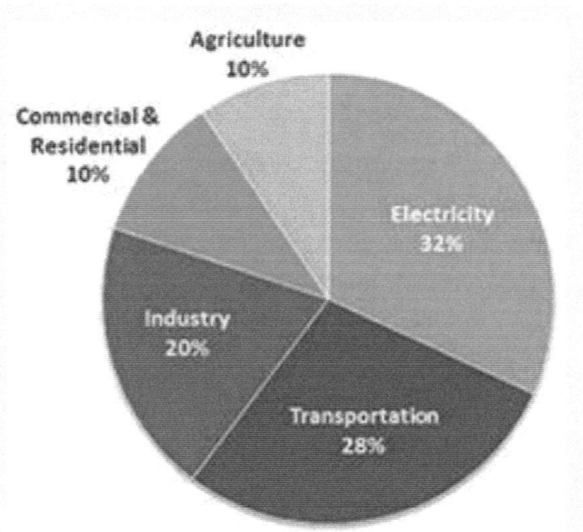

Figure 1. Total Greenhouse Gas Emissions by Economic Sector in 2012 [2]

The implementation of a cleaner energy economy has shown to be a popular policy among voters from both parties. According to a Washington Post – ABC News poll from June 2014, 70 percent of voters say that the federal government should require limits to greenhouse gases from existing power plants. 57 percent of Republicans, 76 percent of Independents, and 79 percent of Democrats support state-level limits on greenhouse gas emissions. Most significantly, voters are willing to pay extra for energy if it means a reduction in greenhouse gas emissions. "Asked whether Washington should still go forward with limits [on energy-sector greenhouse gas emissions] if ‚they significantly lowered greenhouse gases but raised your monthly energy bill by 20 dollars a month,'' 63 percent of respondents say yes, including 51 percent of

Republicans, 64 percent of Independents and 71 percent of Democrats" [3]. These results suggest that Americans on both sides of the aisle are in support of implementing a cleaner energy economy.

In addition to the popularity of a cleaner energy economy, it also makes sense from an environmental and health stand point. On top of the fact that the reduction of greenhouse gas emissions reduces the rate of climate change, it brings along health benefits as well. With the release of its Clean Power Plan in June 2014, which seeks to cut carbon pollution from the power sector by 30% by 2030 by imposing stricter emission limits on existing plants, the EPA projects that "annual public health benefits will total somewhere between $55 billion and $93 billion by avoiding up to 100,000 asthma attacks and 2,100 heart attacks each year." By contrast, they predict compliance cost of $7.3 to $8.8 billion per year, significantly lower than the expected public health benefits [4]

However, despite the ongoing executive branch actions regarding energy policy, the legislative branch has taken relatively little action regarding energy policy lately, mainly due to party-politics-driven gridlock. At the beginning of 2014, the Production Tax Credit for renewables was allowed to expire for the first time since 2004, and only the fourth time since its enactment in 1992 [5]. In May of 2014, the Senate allowed a bipartisan energy efficiency bill to die due to disagreement over other energy related issues, such as the fate of the Keystone XL pipeline and possible legislative response to President Obama"s climate change regulations. In fact, it has been 7 years since Congress passed a major energy bill, which means there has been no major legislative action in response to the changing energy landscape as a result of the shale fracking boom [6].

A plan for implementing a cleaner energy economy becomes even more important as energy demands increase, particularly when it comes to electricity production. As electricity demand increases, more electricity production facilities are built, and how these facilities produce electricity greatly affects their environmental impact. According to the EIA"s Annual Energy Outlook 2014 report, US electricity demand will increase from anywhere between 14% and 41% between 2012 and 2040, though their best estimates suggest that electricity demand will increase by 29% over this time period. Additionally, the growth of electricity demand is projected to increase sometime around 2025, marking the beginning of an upwards slope in electricity demand growth for the first time in 75 years [7].

In order to meet this demand and fill the gap left behind by plant retirements, new electricity production facilities need to be built. As a result of recent emission regulations like the Clean Power Plan, coal power plants are being retired at an accelerated rate. However, due to the low price of natural gas in recent years thanks to fracking, much of this gap has been filled by natural gas fired plants instead of alternatives, like nuclear or renewables. This is mainly due to economics: even the cheapest alternative energy sources, which are on the verge of being if not already cost competitive to coal, simply cannot compete with the low price of natural gas. While natural gas is a lower-emissions alternative to coal, it has emissions nowhere near as low as alternative electricity sources.

The onset of natural gas is not a bad thing, however: using natural gas as a bridge fuel to reduce emissions in the immediate future while hoping that renewables become more cost competitive in the future is not unreasonable given the rate of development and investment in renewable technologies. Natural gas also offers the unique advantage of "spinning reserve", a source that can quickly be powered up to meet rapid fluctuations in demand. This means natural gas can offer a flexibility that base load electricity sources like coal or nuclear cannot, while acting as a buffer against the variability of renewables like solar or wind.

Embracing natural gas in the here and now is also economically responsible, as it keeps electricity prices low by offering a cheap alternative to coal and helps support the growing American natural gas production market. However, without government action, this shift towards natural gas will dwarf any efforts to make renewables a larger share of our electricity production sector. If the government is serious about using natural gas as a bridge fuel to a cleaner energy economy in the electricity sector, a long-term policy needs to be enacted to slowly guide the market in the right direction.

2.0 Overview of Electricity Production Methods and Their Environmental Impacts

In order to discuss energy policy, it is first important to establish the electricity needs of the country, as well as the current capabilities of various electricity production methods and their effects of the environment. As seen in Figure 2 below, coal and natural gas currently dominate the electricity production sector, with nuclear and renewables making up the bulk of the remaining production.

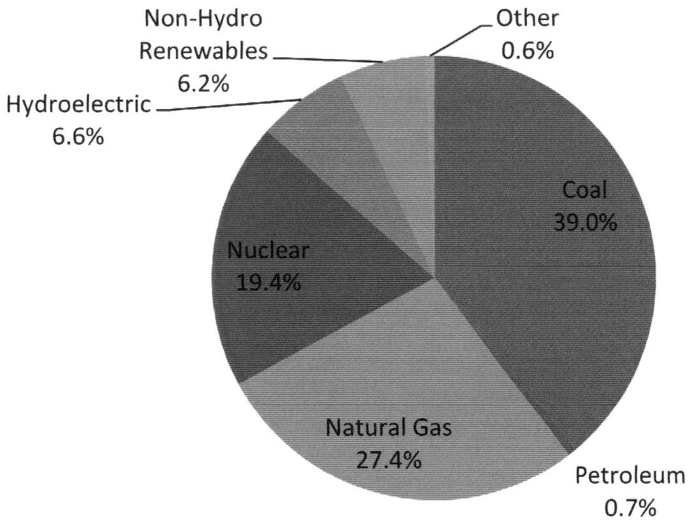

Figure 2. Share of Total US Electricity Production by Source [Adapted from 8]

2.1 Coal and Natural Gas

The two main fossil fuel based / conventional energy producing sources, coal and natural gas power plants are currently the largest source of electricity in America, accounting for nearly two-thirds of the country"s total electricity production [8]. Coal plants are relatively easy to keep in operation over a long time scale, but it is difficult for them to make relatively quick changes to production capacity in response to the changing needs of the grid, particularly during peak-usage hours. As a result, coal plants are very useful in providing a baseline power supply somewhere between peak and off-peak requirements.

Natural gas, on the other hand, allows for a very flexible energy production method. Gas flow rates into a furnace can be controlled with relative precision, unlike solid input streams found in coal plants. As a result, natural gas plants can be very responsive to the changing needs of the energy grid. In fact, natural gas plants are typically used to augment other energy production facilities for this very reason.

From an environmental perspective, the EIA considers conventional electricity sources like coal and natural gas as being the only contributors to CO_2 emissions in the electricity production sector. In 2010, the EIA reported that 1969 million metric tons of CO_2 were released by coal plants nation-wide, while 1285 million metric tons of CO_2 were release by natural gas plants [9].

However, coal tends to have other unique environmental impacts. First, burning coal tends to release nitrogen oxides, which tend to have larger warming potentials than CO_2. For example, N_2O has a SAR 100 year warming potential of 310, meaning that one ton of N_2O has 310 times the impact on global warming compared to one ton of CO_2 over the course of 100 years [10]. From a life cycle assessment standpoint, coal plants produce 974 metric tons of CO_2 equivalent per gigawatt-hour, which accounts for this increased warming potential as well as CO_2 emissions due to non-production factors, such as plant construction or fuel transportation [11]. Coal also tends to produce SO_2, which, along with nitrogen oxides, is the major man-made cause of acid rain. According to the EPA, "In the United States, roughly two-thirds of all SO_2 and one-quarter of all nitrogen oxides come from electric power generation that relies on burning fossil fuels, like coal" [12]. Additionally, coal plants are a major source of mercury, a highly toxic element. According to the EPA, "Coal-burning power plants are the largest human-caused source of mercury emissions to the air in the United States, accounting for over 50 percent of all domestic human-caused mercury emissions" [13].

In the case of natural gas, environmental impacts are lower than that of coal. According to the EPA, "… the burning of natural gas produces nitrogen oxides and carbon dioxides, but in lower quantities than burning coal". In fact, from a life cycle assessment standpoint, natural gas plants produce 469 metric tons of CO_2 equivalent per gigawatt-hour, less than half that of coal [11]. Additionally, the EPA states that "[e]missions of SO_2 and mercury compounds from burning natural gas are negligible". [14]. While the EPA also mentions that unburned natural gas being released either from a power plant or during transportation is a concern, though in 2009 the

EIA stated that "[i]ncreases in energy-related methane emissions [were] largely from underground coal mining", suggesting that methane releases are more prevalent in the procurement of coal than in the use and transport of natural gas [14] [15].

2.2 Nuclear

In the US, nuclear energy makes up nearly one-fifth of all electricity production [8]. According to the World Nuclear Association, "it [nuclear power] is especially suitable for large-scale, continuous electricity demands, which requires reliability (i.e. base load)" [16]. In this way, nuclear power plants are analogous to coal plants in the sense that they are very useful in meeting baseline power needs, but have trouble making quick changes in response to fluctuating peak-hour demand.

From an environmental standpoint, "nuclear power plants do not emit CO_2, SO_2, or nitrogen oxides as part of the power generation process" according to the EPA [17]. However, as with all power production facilities, the construction of nuclear power plants leads to non-negligible greenhouse gas emissions. Based off a compilation of life-cycle emissions studies gathered by the Nuclear Energy Institute (NEI), nuclear energy"s life-cycle emissions are significantly lower than those of coal or natural gas plants, and are in fact comparable to those of renewable like solar, wind, or hydro [18]. From a life cycle assessment standpoint, nuclear produces 15 metric tons of CO_2 equivalent per gigawatt-hour, which is the second-lowest behind wind, and only about 1.6% that of coal plants [11]. Much of these emissions come from the construction of plants, which are significantly larger than other electricity production facilities, and the transportation of fuel.

2.3 Solar and Wind

As shown in Figure 2, only about 6% of electricity production in the US comes from non-hydro renewable sources. Of this 6%, a majority comes from solar (in the form of both photovoltaic cells and solar thermal) and wind [8]. Of all of the available technologies, solar and wind are the most challenging to integrate into the grid due to the fact that they are considered variable sources: if the sun doesn"t shine or the wind doesn"t blow, these technologies are not actively putting energy into the grid, meaning demand has to be met by some other means [19] [20]. In order to offset this, large-scale production facilities are looking into using some form of energy storage system. This would allow excess energy collected during times of sunshine or wind gusts to be used when the sun doesn"t shine or the wind doesn"t blow [19]. However, the storage capacity of current electricity storage technology is limited. As a result, any useful amount of electricity storage is currently too expensive, which limits the viability of renewables.

From an environmental standpoint, neither solar nor wind produce carbon emissions during their operating lifetime, since their energy is derived from the sun and not from burning fossil fuels [19] [20]. However, like nuclear, solar and wind do have some emission impacts due to the processes used to manufacture materials and construct plants: wind is responsible for 14 metrics tons of CO_2-equivalent per gigawatt-hour, while solar photovoltaics are responsible for

39 metric tons of CO_2 equivalent per gigawatt-hour [11]. As with nuclear plants, many of the emissions related to renewables come from construction of production facilities. In the case of solar photovoltaics, higher life-cycle emissions are seen as a result of the emissions that result from the manufacturing process of the solar cells themselves.

2.4 Other Renewables

Solar and wind are not the only renewable sources of energy available to the US. As seen in Figure 2, nearly 6.6% of electricity production comes from hydro plants. The main issue with hydro is that most of the available sites for dams in the US have already been dammed, meaning there is little room for expansion. As a result, the Obama administration has identified a collection of dams that currently do not produce electricity, but outfitting these dams with turbines would only increase the current US hydro capacity by roughly 15% [21]. Another alternative energy source is geothermal, which utilizes naturally-occurring underground sources of heat to power turbines. While this energy is clean, it is only available on a regional basis: Hawaii and the western US have a large geothermal production potential, while areas like New England do not [22]. Since hydropower does not have a lot of potential for growth, and geothermal is regionally specific, they will not be explored as in-depth in this paper.

2.5 Environmental Impact Summary Table

Table 1 below summarizes the life-cycle environmental impact per gigawatt-hour of the main electricity production methods highlighted in this paper.

Technology	Metric Tons of CO_2 released / GWh
Coal	974
Natural Gas	469
Nuclear	15
Solar	39
Wind	14

Table 1. Life-Cycle Environmental Impact Summary of Electricity Production Technologies [Adapted from 11]

3.0 Economics in the Electricity Production Sector

On top of technological limitations and environmental concerns, economics plays a major role in determining which technologies get used for energy production. On the production side, economic factors tend to play a major role in determining which electricity production technologies are utilized by producers. On the consumer side, the electricity production technology used by electricity providers plays a role in the price of electricity sold to consumers.

3.1 Comparing Costs of Electricity Production Technologies

Due the wide range of variables present in energy production, it can be difficult to compare the cost of one electricity production technology to another. However, a useful metric, levelized cost of electricity (LCOE), lays out the cost of given technologies on a per-MWh basis, factoring in such costs as construction, fuel, operating costs, maintenance, and financing costs [23]. Figure 3 below shows the US average LCOE for plants entering service in 2019, in 2012 dollars/MWh, and does not factor in any public financial support (tax credits, subsidies, etc.).

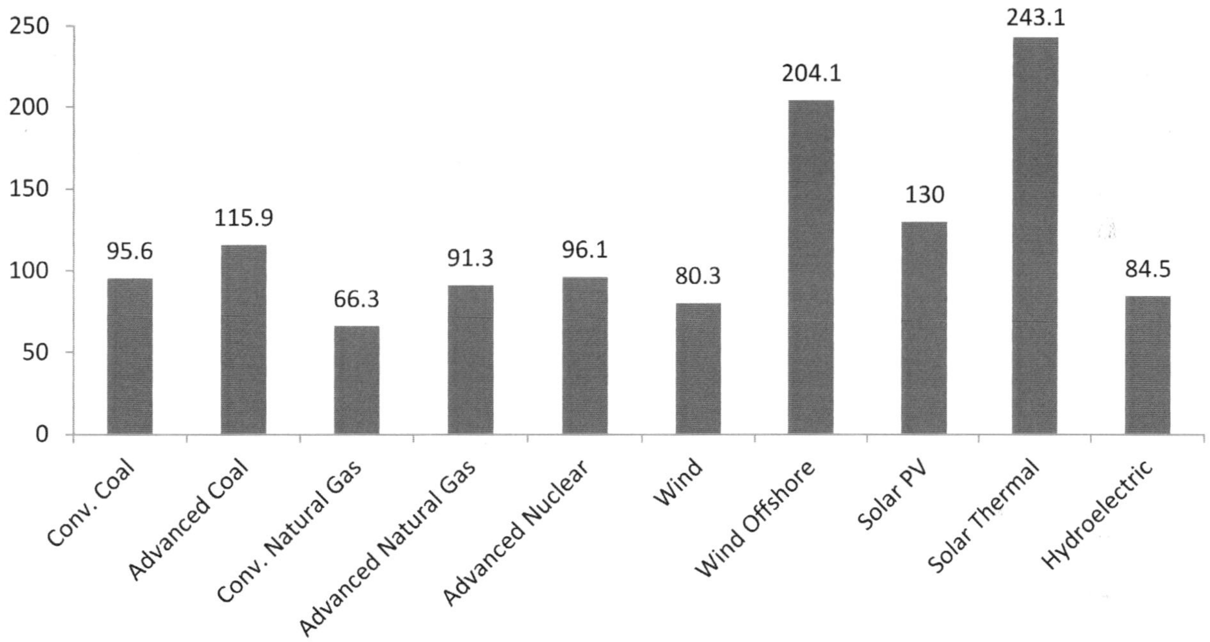

Figure 3. US Average LCOE for Plants Enetering Service in 2019, in 2012$/MWh [23]

As Figure 3 shows, the conventional energy production sources, coal and natural gas, are currently the cheapest available electricity production technologies, which explains why they currently account for such a large portion of US electricity production. Much of this can be contributed to the low price of fuel. Natural gas in particular is benefitting greatly from low fuel prices as a result of the recent American shale fracking boom. However, advanced coal and natural gas facilities, which incorporate carbon capture and sequestration technologies, are significantly more expensive than their conventional counterparts.

While coal and natural gas remain the cheapest electricity production technologies, it is important to note that there are alternatives that are already competitive. For example, the LCOE of nuclear energy is nearly the same as conventional coal and advanced natural gas, and is in fact less expensive than advanced coal. This is particularly noteworthy since nuclear fills the same base load electricity production role as coal, and produces no greenhouse gas emissions during its operational lifetime. It is also important to note that, while many of the renewable energy sources are significantly more expensive than conventional sources or nuclear, wind built on land has the second lowest LCOE, higher than only conventional natural gas.

3.2 Capacity Factor

One factor contributing to the relative expense of renewable sources, such as offshore wind or solar, is their variable nature. Since they are not able to produce electricity at all times of the day, renewable sources suffer from prolonged periods of not operating at peak capacity. The easiest way of quantifying this is through capacity factors, the ratio of actual output to the theoretical output if the plant could operate at full capacity indefinitely. A comparison of capacity factors for each electricity production technology can be seen in Figure 4 below.

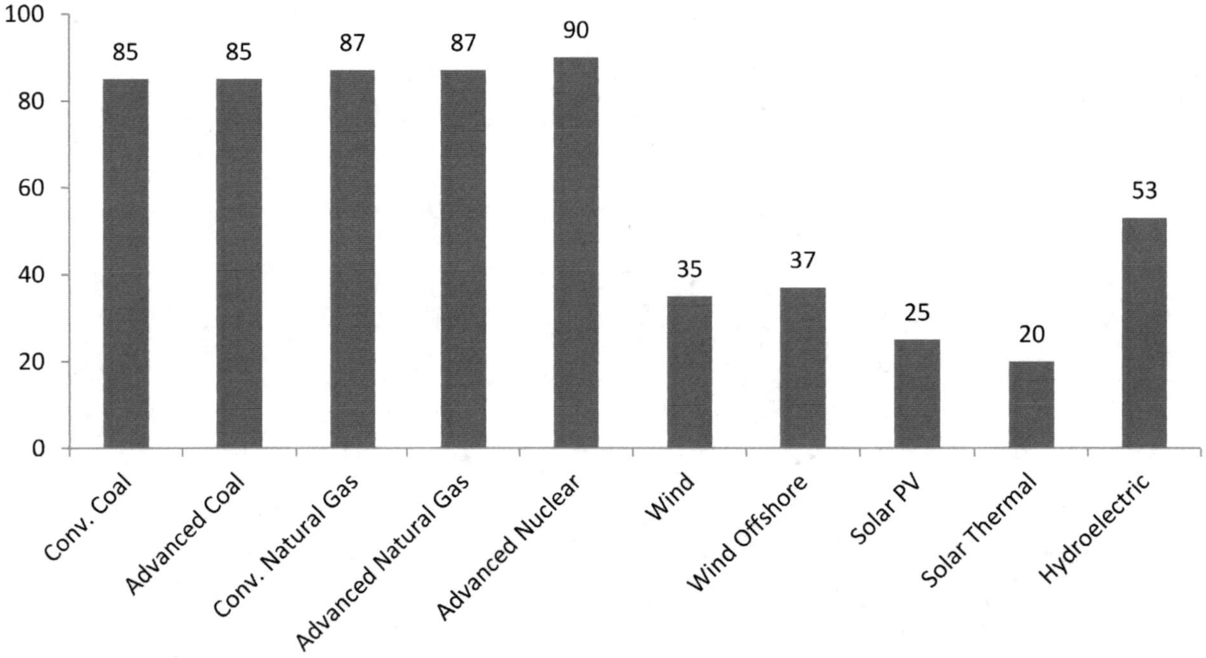

Figure 4. Capacity Factor (%) of Various Electricity Production Technologies [23]

It is important to note that there is no electricity production technology with a capacity factor of 100%: plants do not always run at maximum capacity, and require downtime for things like routine maintenance or system upgrades. However, it is easy to see that renewable sources have significantly lower capacity factors, mainly due to the fact that they cannot produce electricity at all times of the day. Because of this, capital and operating costs of renewable energy plants factor significantly more into their LCOE.

However, with advancements in technology, particularly in the energy storage field, the capacity of renewable sources could significantly increase, and possibly even become comparable with conventional sources. This would cut LCOE by anywhere from 40-70% (depending on current capacity factor). If this electricity storage is cheap enough, even the most currently expensive renewable energy technologies would be competitive with conventional sources.

4.0 Recent Policy Strategies

In recent years, legislators have used a wide variety of strategies to help shape the electricity production sector, most notably by offering some form of economic incentive to desirable technologies. These economic incentives typically come in the form of tax credits, though they may also come in the form of subsidies or guaranteed loan programs. Congress has also made efforts in the past to enact a carbon tax program, but such a program has never been able to make it out of Congress [24] [25].

4.1 Tax Credits for the Electricity Production Sector

The major economic incentive congress has implemented is by extending tax credits to electricity producers, which reduces the amount of taxes they have to pay to the federal government and thus improving their bottom line[1]. Recently, these tax credits have been targeted at renewable electricity resources, most notably solar and wind, though to a lesser extent hydroelectric and geothermal as well [26]. These tax credits have taken on two different forms: an investment tax credit and a production tax credit.

The investment tax credit allows qualifying electricity producers to receive a tax credit equal to a certain percentage of their initial cost of implementation. Solar energy installments are eligible for a 30% credit, while geothermal plants are eligible for a 10% credit. These credits have no limit. As of right now, these tax credits extend to any qualifying facility put in service before January 1, 2017. Wind power facilities put in service between 2009 and 2012 were also eligible for a 30% investment credit, but that has not been renewed [26].

The production tax credit, on the other hand, allows qualifying electricity producers to receive a tax credit based off the amount of electricity produced. As of the end of 2013, this tax credit was 2.3 cents / kWh, and extended to wind, geothermal, and hydropower facilities [5] [26]. If a facility qualifies for both the investment and production tax credit, they are only permitted to claim one of them, at their choosing. The production tax credit extended to any qualifying facility under construction by January 1, 2014, and is available for qualifying facilities for the first 10 years of operation. However, since January 1, 2014, the production tax credit for renewable electricity production facilities has not been extended [26].

4.2 Loan Programs

The federal government also offers loan programs for electricity producers. These loan programs differ from tax cuts mainly in the fact that producers are required to pay back their loan; while there is a cost to the federal government, it is only in the risk of borrowers defaulting,

---

[1] It is important to note that non-profits, such as municipal utilities and utility cooperatives, do not pay taxes, meaning that federal tax cuts do not have an effect on their bottom line. While alternative programs to incentivize non-profits do exist, they are not covered in the scope of this paper.

which has historically been low. Under Title XVII of the Energy Policy Act of 2005, two loan programs were outlined that are still used today: section 1703 and section 1705 [27] [28].

Section 1703 provides loans for "a new or significantly improved technology that is NOT a commercial technology" [27]. Qualifying facilities must also be located in the US. In the current state of the electricity production sector, the only facilities that really meet these requirements are related to nuclear energy; the most recent loans are an $8.3 billion loan for two new nuclear power plants in Georgia and a $2 billion loan for a uranium enrichment plant in Idaho [27]. This program is essential to nuclear power plants, as the most prohibitive cost to entering the market for nuclear plants is upfront capital costs. Due to relatively low operating costs once the plant is built, loan repayment is usually not an issue.

Section 1705, on the other hand, focuses on loan programs for renewable energy production facilities. Unlike section 1703 rules, section 1705 has a time limit, and was only available to projects that had begun construction before September 30, 2011. Since then, this project has not been renewed. Section 1705 loans tend to be lower than their section 1703 counterparts: one of the more expensive section 1705 loans was for $1.3 billion to an Oregon-based wind farm [28].

The major significant difference between section 1703 and 1705 loans is how the Credit Subsidy Cost (CSC) is handled. By law, the CSC, which is "the expected long-term liability to the Federal Government by issuing the loan guarantee", must be paid either by budget appropriations or by the borrower directly. For section 1703, which tend to have more expensive loans and thus larger CSCs, the CSC is paid by the borrower. For section 1705, the CSC was paid for by appropriations [27] [28]. In the American Recovery and Reinvestment Act of 2009, $6 billion was appropriated towards section 1705 CSCs. [29]

4.3 Carbon Tax and Cap-and-Trade Program

On top of offering economic benefits to desirable electricity production methods, Congress can also place economic burdens on undesirable electricity production methods. In the electricity production sector, this can be practically done in one of two ways: either a carbon tax or a cap-trade-program.

A carbon tax works very simply: electricity producers pay an additional tax for each unit of carbon produced during electricity production. This means that carbon-heavy electricity production facilities like coal (and to a lesser extent, natural gas) power plants would pay an additional amount for their electricity production, thus increasing their LCOE. In the same way that tax credits and subsidies seek to level the economic playing field by making renewables cheaper, a carbon tax would seek to level the playing field by increasing the cost of conventional electricity production methods. However, there has never been a carbon tax approved by either chamber of Congress.

An alternative method to a carbon tax is a cap-and-trade program, which seeks to place an artificial limit on the amount of carbon that can be emitted. Once this limit has been determined, electricity producers can buy permits to emit certain amounts of carbon dioxide. The government only sells as many permits as it takes to reach the cap, and then monitors the emissions of each producer to make sure they do not exceed their permitted limit. Producers are able to buy or sell permits to or from other producers as needed.

Cap-and-trade programs create economic incentive to reduce carbon emissions in two ways: first, reducing carbon emissions means that electricity producers need to buy fewer permits, saving them money. Second, if producers reach their permitted cap, their competitors determine the cost of additional permits (since each company now control its own excess permits, not the government). This means exceeding the permitted carbon emission limit can be disproportionately costly, either through the purchase of competitor permits or paying government fines.

In order to ensure long-term carbon emission reduction, cap-and-trade programs also incorporate a permit allowance reduction system. This means that, each year, the amount of permits allowed is reduced, effectively lowering the acceptable amount of carbon dioxide that can be emitted by electricity producers. While a cap-and-trade program did make it out of the House in 2009, the bill was never heard by the Senate, and died when that session of Congress ended [24] [25].

5.0 Projected Electricity Sector Growth: Electricity Sector Capacity Additions for 2013

In order to establish a reference case, it is useful to look at how the electricity production sector has changed recently. Figure 5 below shows the EIA projections for electricity sector capacity additions from 2008.

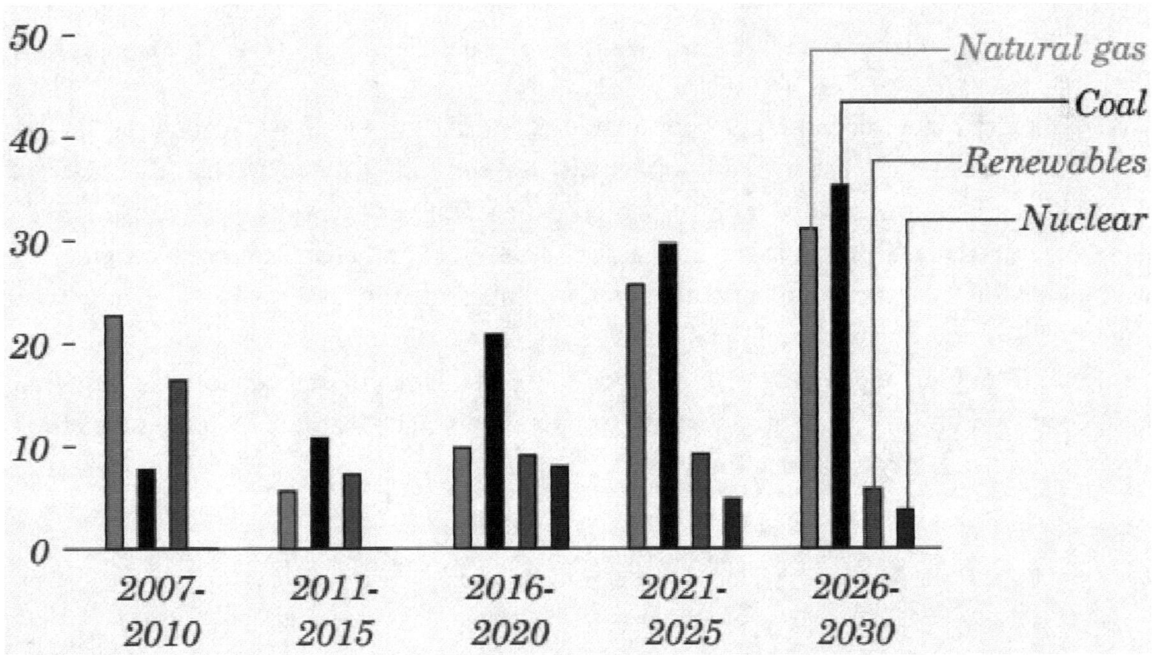

Figure 5. Electricity Generation Capacity Addition Projections by Fuel Type, 2008 (Gw) [30]

As Figure 5 shows, in 2008, it was projected that coal would begin to dominate electricity production additions starting in 2011. Much of this can be attributed to the increasing cost of natural gas, as shale fracking was still in its infancy in 2008. Additionally, renewables, which see a large spike in 2007-2010, were projected to fall by the wayside after 2010 as they begin to be dwarfed by both natural gas and coal.

However, as a result of the shale fracking boom, the reduced cost of renewables, and increased government pressure to reduce carbon emissions, these 2008 projections do not reflect what has actually happened in recent years. Figure 6 below shows the electricity generation capacity additions for 2013.

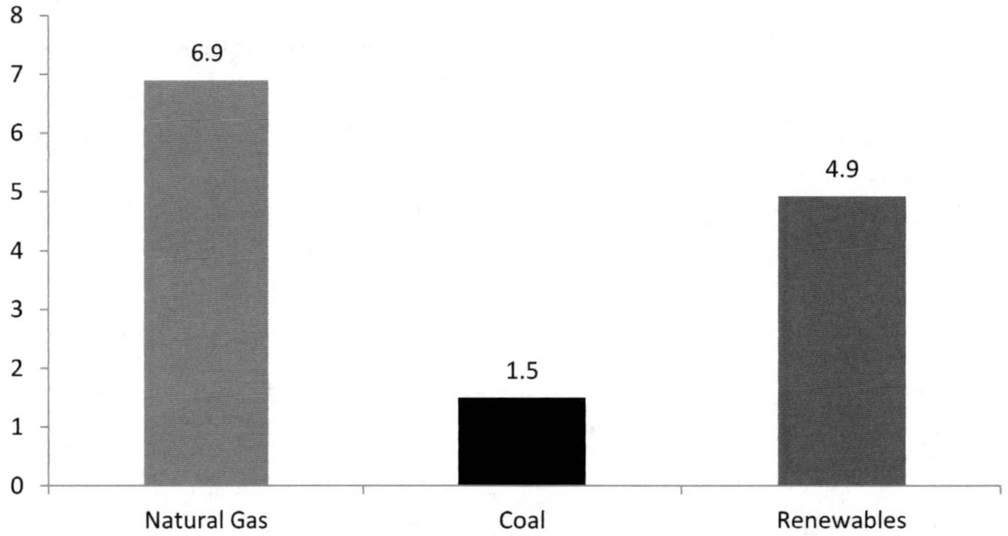

Figure 6. Electricity Generation Capacity Addition by Fuel Type 2013 (GW) [Adapted from 31]

As seen in Figure 6, natural gas dominated new capacity additions, accounting for more than half of the electricity capacity additions in 2013. Renewables made up roughly one-third of new additions, which is attributed to the completion of a number of large scale solar thermal plants in Arizona and California as well as “falling [photovoltaic] costs, aggressive state renewable portfolio standards, and continued federal investment tax credits” [31]. Coal makes up a meager 11% of new capacity additions, in stark contrast to the projections shown in Figure 5.

Long term projections have also changed to reflect the changing economic and political landscape surrounding electricity production. Figure 7 below shows the EIA projections for electricity sector capacity additions from 2014.

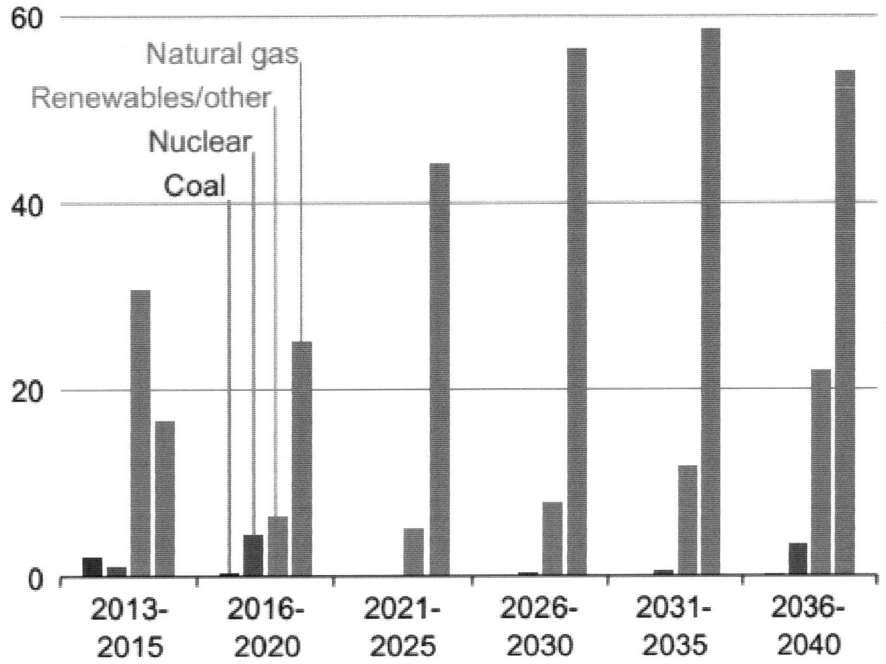

Figure 7. Electricity Generation Addition Projections by Fuel Type 2014 (GW) [32]

As Figure 7 shows, natural gas dominates new additions through 2040, mainly as a result of low gas prices. Coal additions fall virtually to 0 as a result of increased emissions regulations. Most alarmingly, the addition of renewables is actually lower than those projected in 2008 for the period between 2016 and 2035. This is mainly due to the extremely low cost of natural gas relative to any other fuel source, making it take up a larger share of the new additions in the 2014 projections than conventional sources did in the 2008 projections. It is important to note, however, that these projections only show a reference case where no additional government actions are taken.

5.1 Policy Action Effects: Overall Carbon Emissions

Despite the recent transition from coal to natural gas, as well as the increased onset of recently-cheap renewables, carbon emissions from the electricity production sector are projected to increase over the next several decades if current laws and regulations remain in effect. This is mainly attributed to the shift towards natural gas-based electricity production and the shift away from renewables and nuclear as a result of cheap natural gas for the foreseeable future [33]. However, if some form of carbon tax is implemented, carbon emissions could decrease instead increase, as seen in Figure 8 below.

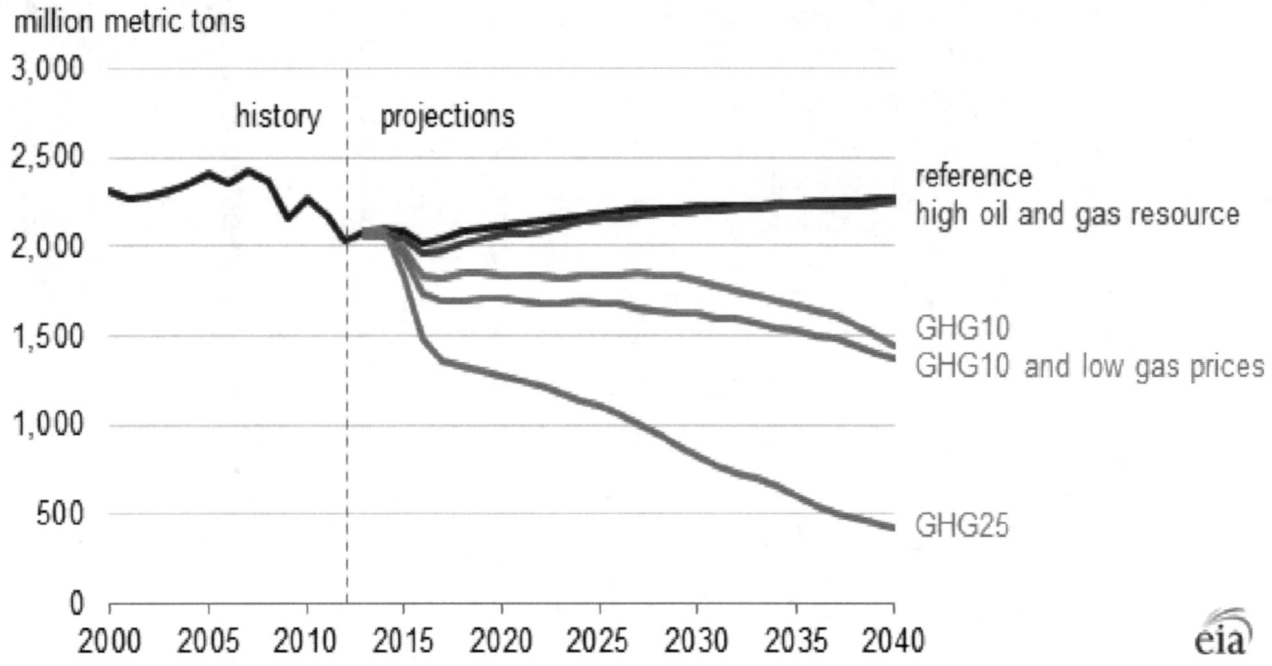

Figure 8. Electricity-Related CO_2 Emissions in Five Cases, 2000-40 [33]

Figure 8 takes into account five different potential cases. The reference case reflects no change in laws or regulations surrounding electricity production and assumes relatively constant economic factors relative to those in 2014. As seen in Figure 8, the reference case shows a 12% increase in emissions up through 2040 [33].

The GHG 10 and GHG 25 cases both refer to the implementation of a carbon tax (or equivalent cap-and-trade permit cost) in 2015 of $10 / metric ton of CO_2 or $25 / metric ton of CO_2, respectively. The GHG 10 case shows a 16% lower CO_2 emission rate in 2025 and a 36% lower CO_2 emission rate in 2040 relative to the reference case. The GHG 25 case shows an even greater decrease in CO_2 emission rates, with a 49% reduction in 2025 and an 82% reduction in 2040. In both cases, there is a large dip in emission rates before 2020, partially due to an accelerated shift away from coal and partially due to an increased implementation of renewables [33].

The High Oil and Gas Resource case assumes the same conditions as the reference case, except that oil and gas resources become more abundant than currently projected. In this case, CO_2 emissions decrease ever so slightly more than in the reference case as a result of increased shift towards natural gas, but by 2040 emission rates are roughly the same as the reference case as a result of producers expressing and increased favor towards natural gas relative to renewables [33].

Finally, the GHG 10 and Low Gas Prices case effectively combines the GHG 10 and High Oil and Gas Resources cases. In this case, reduction of CO_2 emissions before 2020 is greater than those seen in the GHG 10 case, mainly due to an even greater shift from coal to natural gas. However, because a greater share of this shift away from coal is given to natural gas and not renewables, CO_2 emissions do not decrease as drastically after 2030 compared to the

GHG 10 case; it can be extrapolated that CO_2 emissions would actually begin to be higher than the GHG 10 case sometime between 2040 and 2050 [33].

Overall, these cases suggest that some form of action needs to be taken in order to produce a marked reduction in carbon emissions from the electricity production sector. If nothing changes, CO_2 emissions will increase slightly over the next 40 years, in contrast to the significant emission reductions we‘‘ve seen since 2005. Emissions get even worse with an increased gas resource and lower gas prices, which may result with the advancement of fracking and other extraction technologies. Based off of this projection, the only way to continue a trend of CO_2 emission reduction is to accelerate the shift away from coal and ensure that an increased share of that shift is made up of renewables or nuclear.

5.2 Policy Action Effects: Conventional Fuels

When assessing the viability of the cases listed in the previous section, it is important to understand the effects of each action on each specific electricity production method. As discussed previously, there is already a shift towards natural gas and away from coal in response to lower gas prices and increased carbon emission regulations. Figure 9 below shows how possible policy actions may augment that shift by showing the projected natural gas electricity generation in the five previous cases.

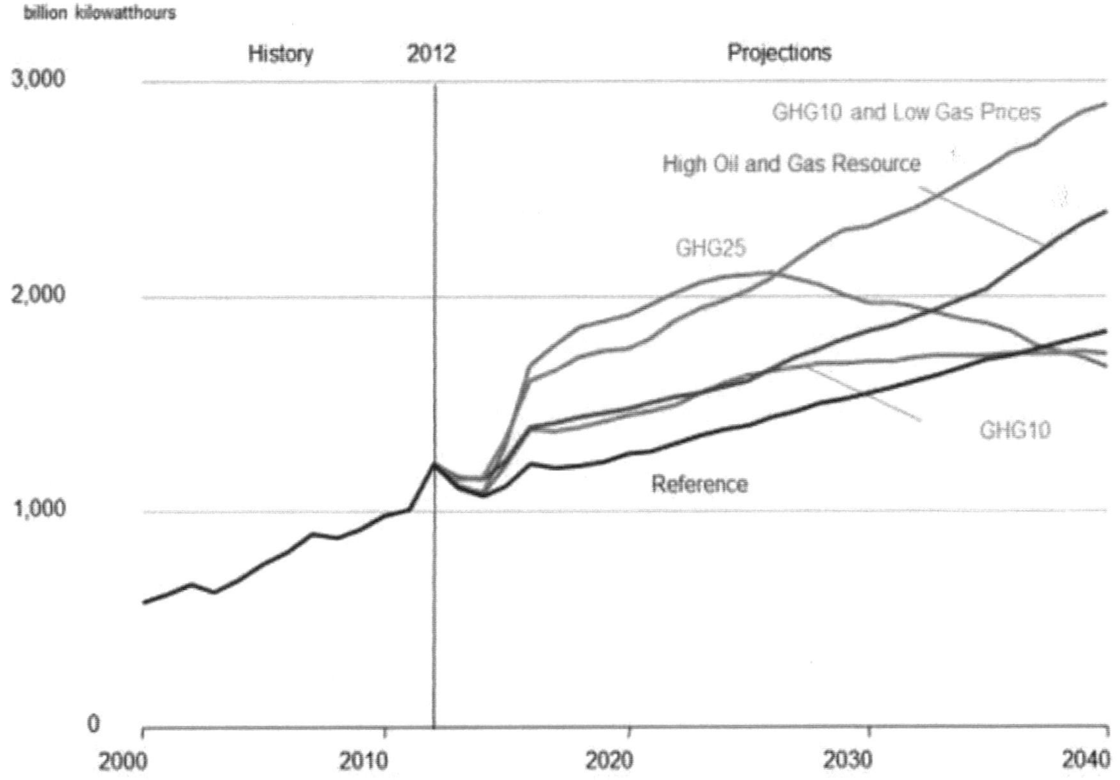

Figure 9. Natural Gas Electricity Generation in Five Cases, 2000-40 [34]

As seen in Figure 9 above, electricity production via natural gas is projected to increase over the next several decades in all cases. In the reference case, we see a relatively stable and constant increase of natural gas fired plants as they replace retiring coal plants; in 2012, 37% of electricity generation comes from coal, and in 2040 that share is 32%, marking a relatively slow shift [34]. With a high gas resource, this shift is more dramatic in the near future, and the long term rate of increase in natural gas plants is also slightly higher than the reference case.

The most interesting part of this projection, however, is the effects of a carbon tax. In all GHG cases, the shift towards natural gas in the near future is increased relative to the reference case. In the GHG 10 case, the share of electricity from coal is 28% in 2025 and 19% in 2040. For the GHG 25 case, the shift is even more drastic, with the coal share being 10% in 2025 and 1% in 2040 [34]. In both cases, the increase in natural gas production is curbed by the increased implementation of nuclear of renewables: in the GHG 10 case, natural gas electricity production levels off around 2030, while in the GHG 25 case, it actually decreases starting in 2025. However, if gas prices remain low, natural gas electricity production continues to increase through 2040, reaching the highest level of any of the five test cases.

5.3 Policy Action Effects: Renewables

Despite the fact that natural gas dominates new electricity production additions, renewables are also seeing an increase in implementation as the electricity production shifts away from coal. However, without some form of government action, the rate of renewable additions will be dwarfed by those of natural gas, leading to the increased carbon emissions seen in Figure 8. Figure 10 below shows the effects of some possible government actions by plotting the projected renewable electricity generation in four cases.

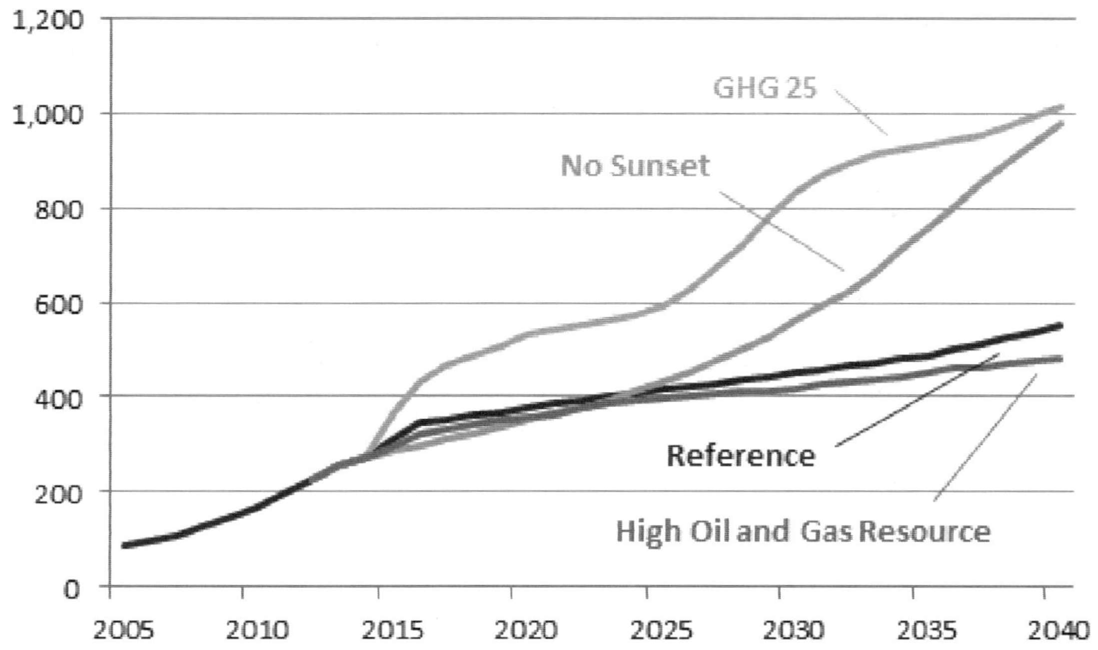

Figure 10. Renewable Electricity Generation in Four Cases, 2000-40 (billion kWh) [Adapted from 35]

In all cases, electricity generation from renewables increases over the next few decades. However, for the reference and high gas resource cases, the rate of increase in renewables is less than that seen between 2000 and 2014, mainly due to the increased implementation of natural gas power plants. The no sunset case shows the effects of extending renewable tax credits (i.e. the installment tax credit and the production tax credit) indefinitely in their current forms. In this case, the growth seen between 2000 and 2014 remains relatively constant through 2025, at which point the rate of implementation of renewables begins to increase steadily. In the GHG 25 case, a large increase is seen just after 2015 in response to the implementation of the carbon tax, after which the rate of implementation of renewables begins to level off. However, around 2025, electricity generation from renewables begins to increase dramatically once again, as more projects come into completion in response to the increased cost of natural gas power plants, whose production begins to decline as seen in Figure 9.

Compared to the reference case, both the no sunset and GHG 25 cases see disproportionate increases in solar and wind relative to other renewables like hydro or geothermal. When it comes to solar, electricity production in both cases is triple that of the reference case in 2040; for wind, it‘‘s double. In both cases, geothermal actually sees slightly lower electricity production in 2040 compared to the reference case, and hydro sees relatively little growth between 2014 and 2040 in all cases [35].

6.0 Policy Recommendations

The electricity production sector in the US is rapidly changing as a result of the shale fracking boom. While the large shift from coal to natural gas will improve carbon emissions on a per kwh basis, the low price of natural gas is also reducing the amount of renewable sources entering the market. If no government action is taken, the growth of electricity demand will still mean an increase in carbon emissions from the electricity production sector over the next 30 years due to the low prices of natural gas. If the government truly wishes to use natural gas as a bridge fuel to a low-carbon electricity sector, a carbon tax needs to be implemented and tax credits for renewables need to be extended.

A carbon tax is necessary to appropriately reflect the societal cost of carbon-emissions-heavy technology. However, a large carbon tax is not necessary, and would cause unnecessary instability as the market acted swiftly to adjust. Instead, a low carbon tax that progressively increases is the most responsible way to implement such a change. Based off the projections in Figures 8 and 9, a carbon tax of $10 / metric ton of CO_2 would lower carbon emissions in the electricity production sector without drastically changing the economic viability of natural gas facilities. This tax would increase at a rate of 5% per year in order to incentivize investment in further carbon emission reduction technology. For a cap-and-trade program equivalent, this means the acceptable level of carbon emissions each year would be reduced, effectively increasing the cost of emissions a proportional amount.

On the other side of things, renewable technology still needs some assistance to be competitive with conventional fuels, as seen in Figure 3. However, the current split between the investment tax credit and the production tax credit makes the development of new and improved renewable technologies difficult, as it is more difficult to predict the economic environment these technologies will emerge into. A single, long-term tax strategy would offer economic stability and predictability that promotes private investment that has been yet unseen. In this way, a single, long-term tax strategy should be used to incentivize the use of all types of renewables in the electricity sector.

Of the two programs currently available, fully embracing the production tax credit makes the most sense, as it incentivizes the implementation of energy efficient and energy dense technologies[2]. It also incentivizes the development of improvements that reduce the upfront cost of these technologies, which help open them up to small scale residential and commercial consumers. The current rate of 2.3 cents per kwh is an appropriate start-point for such a tax program, but with the implementation of a carbon tax, indirect incentive towards renewables increases proportionately. The increasing nature of the carbon tax combined with the increased rate of improvement for renewable technologies actually means that the production tax credit can be slowly reduced. As such, a 5% reduction per year to mirror the carbon tax rate increase would be appropriate.

In both the carbon tax and the production tax credit, both programs need to be comprehensive and long term in order to allow the electricity producers to accurately model the economic environment that they will be entering with new projects. Due to the long-nature of construction of electricity production facilities, as well as the long operating life-time of these

---

[2] Again, this credit would not incentivize non-profit utility providers. Alternative programs, not covered in the scope of this paper, could be used to incentivize these utility providers.

facilities, stability in the tax structure from the government is paramount. The unexpected expiration of the production tax credit in 2013 mainly to political gridlock is unacceptable, and should be avoided in the future if at all possible.

All in all, if such a two pronged approach of a $10 / metric ton carbon tax combined with an extension of the production tax credit is used, the government will turn a net profit as a result. Based on the projections in Figures 9 and 10, there will be an average profit of $5.5 billion per year between 2015 and 2040 under such a program, with yearly profits ranging between $4.5 billion and $6.1 billion [adapted from 40 and 41]. Keep in mind that these projections only factor in the revenue from natural gas facilities; factoring in the revenue of coal should increase revenue further, despite the increase in cost from the increased use of renewables.

It is further recommended that the money made from this program be reinvested in the loan programs for electricity producers. While the reintroduction of the section 1705 program may be beneficial, it seems more beneficial to use this money to offset or eliminate the CSC cost burden on borrowers of section 1703 loans. With the accelerated retirement of coal, reduced or eliminated CSC costs would incentivize increased use of nuclear power, which acts as a more practical and carbon neutral base load power supply alternative compared to natural gas.

7.0 References

[1] Obama, B. (2014, January 28). *President Barack Obama's State of the Union Address.* Speech presented at the House Chamber, Capital Building, Washington, DC. Retrieved from http://www.whitehouse.gov

[2] EPA. (2014, June 2). *Learn About Carbon Pollution From Power Plants*. Retrieved from http:// www2.epa.gov

[3] Clement, S. (2014, June 2). A huge majority of American support regulating carbon from power plants. And they're even willing to pay for it. *The Washington Post*. Retrieved from http://www.washingtonpost.com

[4] Koch, W. (2014, June 3). EPA carbon emission limits may shift energy mix. *USA Today*. Retrieved from http://www.usatoday.com

[5] Union of Concerned Scientists. (2014, January 31). Production Tax Credit for Renewable Energy. Retrieved from http://www.ucsusa.org

[6] Davenport, Coral. (2014, May 12). Amid Pipeline and Climate Debate, Energy-Efficiency Bill Is Derailed. *New York Times*. Retrieved from http://www.nytimes.com

[7] EIA. (2014, May 7). *Annual Energy Outlook 2014*. "Market Trends: Electricity demand". Retrieved from http://www.eia.gov

[8] EIA. (2014, May 21). *Electric Power Monthly*. "Table 1.1. Net Generation by Energy Source: Total (All Sectors), 2004 – March 2014". Retrieved from http://www.eia.gov

[9] EIA. (2013, May). *State-Level Energy-Related Carbon Dioxide Emissions, 2000-2010.* "Table 2. 2010 state energy-related carbon dioxide emissions by fuel". Retrieved from http://www.eia.gov

[10] EIA. (2011, February). *Emissions of Greenhouse Gases in the United States 2009*. "Table 5. Greenhouse gases and 100-year net global warming potentials". Retrieved from http://www.eia.gov

[11] Meier, P.J. (2002, August). *Life-Cycle Assessment of Electricity Generation Systems and Applications for Climate Change Policy Analysis*. pg 70. Madison, Wisconsin: University of Wisconsin, Fusion Technology Institute. Retrieved from fti.neep.wisc.edu/pdf/fdm1181.pdf

[12] EPA. (2012, December 4). *What is Acid Rain?*. Retrieved from http://www.epa.gov

[13] EPA. (2014, March 10). *Basic Information | Mercury*. Retrieved from http://www.epa.gov

[14] EPA. (2013, September 25). *Natural Gas | Clean Energy*. Retrieved from http://www.epa.gov

[15] EIA. (2011, March 31). *Emissions of Greenhouse Gases in the US*. Retrieved from http://www.eia.gov

[16] World Nuclear Association. (2014, May). *World Energy Needs and Nuclear Power*. Retrieved from http://www.world-nuclear.org

[17] EPA. (2013, October 22). *Nuclear Energy | Clean Energy*. Retrieved from http://www.epa.gov

[18] Nuclear Energy Institute. (2014). *Life Cycle Emissions Analyses*. Retrieved from http://www.nei.org

[19] EPA. (2012, February 28). *Solar Energy | Energy and Global Climate Change in New England*. Retrieved from http://www.epa.gov

[20] EPA. (2014, May 1). *Wind Energy | Energy and Global Climate Change in New England*. Retrieved from http://www.epa.gov

[21] DOE. (2012, April 17). *Energy Department Report Finds Major Potential to Increase Clean Hydroelectric Power*. Retrieved from http://energy.gov

[22] EPA. (2014, May 1). *Geothermal Energy | Energy and Global Climate Change in New England*. Retrieved from http://www.epa.gov

[23] EIA. (2014, April). *Levelized Cost and Levelized Avoided Cost of New Generation Resources in the Annual Energy Outlook 2014*. Retrieved from http://www.eia.gov

[24] American Clean Energy and Security Act of 2009. H.R. 2454 of the 11th Congress. Retrieved from http://www.opencongress.org/bill/111-h2454/actions_votes

[25] Hulse, Carl. (2010, July 22). Democrats Call Off Climate Effort Bill. *New York Times*. Retrieved from http://www.nytimes.com

[26] 26 U.S.C. §§46, 48

[27] DOE. *Section 1703 Loan Program*. Retrieved from http://energy.gov

[28] DOE. *Section 1705 Loan Program*. Retrieved from http://energy.gov

[29] Wilson Sonsini Goodrich & Rosati. *Status Update – Department of Energy Title XVII Loan Guarantee Program*. Retrieved from http://www.wsgr.com

[30] EIA. (2008, June). *Annual Energy Outlook 2008*. "Early Capacity Additions Use Natural Gas, Coal plants Are Added Later". Retrieved from http://www.eia.gov

[31] EIA. (2014, April 8). *Half of power plant capacity additions in 2013 came from natural gas*. Retrieved from http://www.eia.gov

[32] EIA. (2014, May 7). *Annual Energy Outlook 2014*. "Most new capacity uses natural gas and renewables". Retrieved from http://www.eia.gov

[33] EIA. (2014, June 9). *Power sector CO_2 emissions sensitive to policy changes and natural gas supply*. Retrieved from http://www.eia.gov

[34] EIA. (2014, May 7). *Market Trends: Emissions*. "Energy-related carbon dioxide emissions are sensitive to potential policy changes". Retrieved from http://www.eia.gov

[35] EIA. (2014, April 29). *Renewable electricity projections show growth under alternative assumptions in AEO2014*. Retrieved from http:// www.eia.gov

393632 Turning the Big Idea into a Startup

Abstract:

When a university researcher or garage inventor is thinking about commercializing an invention, there are a number of early considerations that are not always obvious to first time technical entrepreneurs. This talk will focus on some of the early stage commercialization decisions and how those decisions may impact the startup company potential later. Some of the topics to be discussed include fundraising, teambuilding, and intellectual property management.

386885 Starting an Small Engineering Consulting Business – the 10 Year Anniversary of a Startup Company

Abstract:

A small engineering consulting business was started in 2004 with the modest objectives of addressing short-term business opportunities while remaining in a specific geographic location. Full-time commitment to this venture occurred for a seven-year period. Ten years and 154 engineering projects later the business continues to operate. Personal and professional lessons learned on business planning and strategies, accounting and financial tools, fixed asset investments, family involvement, and finally, the pros and cons of this career path versus the traditional non-ownership career path will be discussed.

387532 The Art of the Mass Balance: The Cultural Tools of Entrepreneurial Chemical Startups

Abstract:

The product cycle processes of tech based startups and chemical startups are inherently different. Tech startup thought leaders teach entrepreneurs to hit the ship date even if all the kinks aren't worked out. Later a patch can be uploaded by customers to fix bugs or add functionality at minimal cost to the startup.

Chemical startups differ significantly as processing plants are capital intensive and must incorporate safety, equipment lead times, and construction efforts. Adopting an entrepreneurial approach, chemical startups can deliver against the fast paced timeline while designing the flexibility required to commercialize a first of its kind technology. The tool for success in an entrepreneurial chemical startup is the mass balance.

Keeping score in an industrial startup is dependent upon the value and quality of the product. The disposition of the constituents in the product determines the cash flow possibilities in the operation cost and the needed investment to enable the conversion of the raw material feedstock mass into the valued finished product. The key to a successful mass balance is the validation of the instruments and test methods measure the chemical engineering entrepreneurial effort's success.

The Personal and Family Side of Starting an Independent Business

Starting a business has its special and unique technical challenges relating to deciding on a focus, designing a strategy for communication and outreach, identifying clients, obtaining financing, and many others. But unless you are a single adult, there is another factor and that is your family. If they have gotten used to a regular paycheck, travel being reimbursed, promotions coming on a regular business, and the occasional special recognition, this can all come to a screeching halt. The support and understanding needed from a family and external networks is critical for a new venture to succeed. The level of this support will also depend greatly on how large the imagined enterprise is envisioned to be.

These are some questions to consider:

1. <u>Short term economic survival.</u> How long can you survive without a regular paycheck? Have you built up sufficient savings in the event that your first clients are harder to find than you thought? If your spouse is not working, will they be able to work to cushion the cash flow? Do they want to? What is the market in the local area for the skill set they possess? Will the cost of child care be part of the equation? It is not unusual at all for someone starting an independent business to wait at least six months before receiving their first payment. More companies are moving to 90 day plus payment cycles. Is that tolerable? A discount on pricing and acceptance of a client's corporate credit card are ways of increasing short term cash flow at a small long term cost. There may also be short term training opportunities available through various organizations, but not necessarily at the same rate of pay that may be desired long term.

2. <u>Health insurance.</u> The degree to which this is of concern will depend on a large number of factors including age, state of residence, health history barriers to insurability, and whether a working spouse has insurance to which the new "independent" can be added. Regardless of the specific answers to these questions, the cost of covering health insurance must be considered when starting an independent business. One way of minimizing a major impact is to purchase low cost catastrophic insurance with a very high deductible, usually available in many areas without medical questions. And of course, AIChE offers group health insurance plans, but which come with holding periods before covering pre-existing conditions.

3. <u>Spouses.</u> No two people are created equal and the reaction of family members to such a venture, even when openly discussed ahead of time, can vary and change. What is the personal tolerance for going without a paycheck for an extended period of time (even assuming the points in "short term survival" have been addressed)? Is how the neighbors and your social circle view your situation important? If you are a male and your wife is working or needs to go back to work, are you willing to do household chores you've never done before? Some of these attitudes will depend upon whether your starting of an independent business was totally your choosing or as an indirect result of a reduction in force, a general layoff or closure/relocation of a business, or a personal performance issue. Is your spouse willing to cut back on some of the luxuries they may have enjoyed previously?

4. <u>Children.</u> This is a different issue than spouses and will greatly depend on their ages. What would your child's reaction be to someone in their school, possibly informed by one of your neighbors about your situation, saying something like "Your mommy or daddy is out of work…ha, ha". Will they start a fight? How well have you prepared them?

5. <u>Support structure.</u> In today's world, the stigma of losing a job is far less than it used to be due to numerous factors including mergers and acquisitions, change in business emphasis, and "forced" geographic moves to retain a job. Yes, there are still performance issues, but many times this can be a result of different approaches to a problem or opportunity vs. genuinely poor performance. It is

highly unlikely that anyone today will have only one job for their entire career. Since everyone, over some period of time, will experience the same kind of dislocation, there is virtually no barrier to discussing possible job opportunities with anyone. Today's professional networking tools such as LinkedIn are tremendous sources of connections and opportunities. Local professional engineering chapter meetings are another great source, especially if there is a strong desire to stay in the same geographic area.

It is important to remember that starting such a venture within a family structure has additional needs to consider beyond just hanging out the professional consulting shingle. Consider them all before taking the plunge and your venture has a much higher probability of success.

392019 Aligning Innovation, Patent Strategies, and Business Models

Abstract:

An iterative approach to balancing research efforts, patent strategies, and business models can identify new opportunities as well as provide an attractive package for management or investment support. An overview of the beneficial use of intellectual property is provided and specific examples are shown.

364257 Building an Excellent Entrepreneurial Team

Abstract:

An entrepreneurial team is built on excellence, confidence and passion. All team members must be comfortable with chaos. The cube farm mentality has to be destroyed. The passion to change the world and make a difference has to be unleashed. The recruiting relationship is built on the abnormal. There is no room for conformity on an entrepreneurial team. However, the same traits that make it great also tends to blow it apart.

Entrepreneurial engineering leaders are either superstars or oddballs. Both traits are needed. The team leader must find the mix of innovative off the wall creativity with the ability to drive fast-paced targets and get things done. Decisions must be made on the fly without all the information, but the aspects of safety and ethics can never be compromised. Innovation is unleashed by having mundane tasks taken out of the equation.

An entrepreneurial team is a paradox. Like the grand unified force theory, the moment you think you have handle on how it works, a new opportunity emerges and bang ! pivot to a new direction.

There is nothing more fulfilling, exciting or satisfying.

392020 How to Generate Revenue from Your Technology: Strategies for Licensing Your Intellectual Property

Abstract:

Licensing of intellectual property can be a significant source of revenue for companies. This session will provide practical lessons for generating and increasing value from IP through licensing. Learn how to identify and protect your IP while avoiding common mistakes when monetizing it.

consulting engineers and scientists

GEI Consultants

John T. Finn, P.E.
GEI Consultants, Inc.

Dedication

In Memory of My Father

Robert K. Finn, PhD.
Professor of Chemical Engineering
Cornell University
Member AIChE 1946 - 2012

GEI Consultants

Create Winning Boards for Your Startup

You Want to Be Here

You Are Here

GEI

Purpose ?

- What do you want from outside help?
- Depends on:

1. Unique Identity and Circumstances
2. Stage: Idea ➡ Startup ➡ Growth

GEI

Directors, Advisors, and Supporters

- Firm reports to the Directors

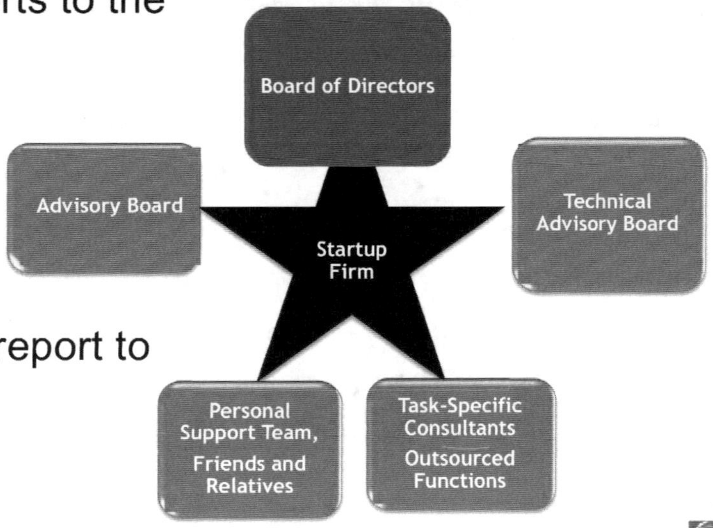

- Advisors report to the Firm

GEI Consultants

Three Types of Boards

Board	Legal Status	Primary Role
Directors	Statutory	Makes Big Decisions, Goals, Selects CEO
Advisory (General)	Informal	Advises on vision, goals, strategy, the market, legal, staffing, financial, governance, etc.
Technical Advisory	Informal	Advises on technical and scientific aspects of the process and product, R&D, external related technologies, etc.

GEI Consultants

Directors

- Usually called "The Board"
- Usually after Idea and Initial Startup
- Founders, VC Investors and Outside Members
- How Many: 7 to 9 to fulfill legally mandated activities
- Commitment: 250 hours/year Chair or active startup 450 hours/ year

GEI Consultants

General Advisors

- Provide critical business-related help, introductions to other key supporters
- Focus on your path to the next stage
- 3 to 5 persons, with a range of backgrounds
- For the number of advisors, consider your own bandwidth. 1 hr each per month, min.

GEI Consultants

Technical Advisory Board

- Keep it small and focus on your challenges
- Academic and Industry backgrounds
- Include your technical team members
- Assign members 10 minute presentations on an aspect of the field
- Encourage constructive dialogue
- Stress confidentiality of your data

GEI Consultants

TAB Special Considerations

- Certainty vs Risk for engineers.
- Audacity (arrogance ?) for Chem Es!
- Humility is the greatest virtue for the TAB

GEI Consultants

Finding Fellow Innovators

- Meet a lot of people. Reach outside your circle, referrals from college profs, grad students, conferences, accelerator meetings, AIChE, LinkedIn

...let people know you are looking.

- Board membership benefits the members: Pitch
- Be picky! Proven track record
- Passion about your startup
- Bandwidth to be able to help
  - test this by giving a small assignment
- Good chemistry with you, and board members
- Collaborative, yet challenge assumptions

GEI Consultants

Getting the Most from Advisors

- Meet both individually and in groups
- Run the meetings efficiently, with pre-distributed agenda, minutes.
- Give them homework, specific tasks
- Set the next meeting date
- Use calls, email and chat effectively
- Consider a secure site Wiki with company updates, financials, tech developments, marketing tools, Suggestion Box, etc.

GEI Consultants

Compensation

Board	Vested Equity	Cash	Meeting Expenses, Lunch
Directors	0.5% to 2% if not a VC or Founder	Not usually	Yes
Advisory (General)	0 to 1%	Small	Usually
Technical Advisory	0 to 1%	Small	Usually

5% Total equity pool set aside for advisors is typical

GEI Consultants

Standard Agreement

- Founder Institute

Founder/Advisor Standard Template "FAST"

- Easily defines

scope, time commitments, compensation

- Proportional to stage: Idea, Start, or Growth
- Term limit, typically 1 year

GEI Consultants

Advisors Compensation

- Wide Range: Lunches at first, to Equity

FAST Scope	Idea Stage	Startup Stage	Growth Stage
Standard: Monthly Meetings	0.25%	0.20%	0.15%
Strategic: Add Recruiting	0.50%	0.40%	0.30%
Expert: Add Contacts & Projects	1.00%	0.80%	0.60%

Source: The Founder Institute

GEI Consultants

Supporters and Consultants

- Asking the Deeper Questions
- Improve Structure
- Project Management

Systems

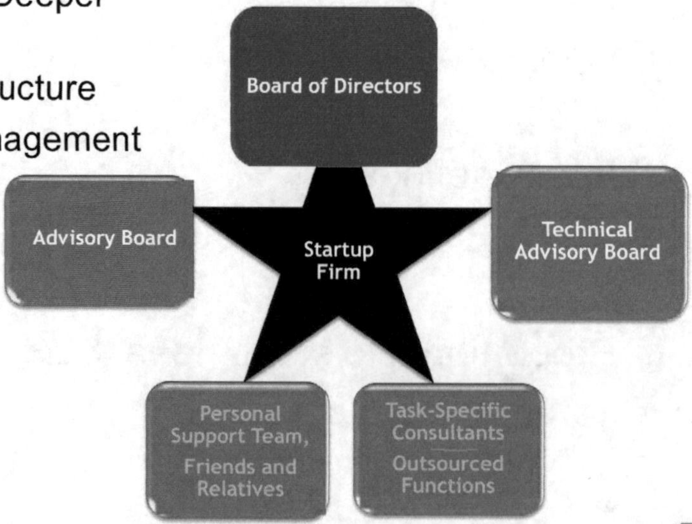

GEI Consultants

Critical Points

- Advisory Boards are essential for most Startups to thrive and progress
- Invest time and attention in setting up and maintaining the board
- Consider general business advisors as well as expert and diverse technical advisors
- Tailor board makeup, scope, compensation to your stage
- Use consultants for specific outsourced tasks
- Friends and family provide personal support

GEI Consultants

Thank You

Questions and Discussion ?

GEI Consultants

Resources

- Entrepreneur article: www.entrepreneur.com/article/71684
- Ventureneer blog: bloghttp://ventureneer.com/vblog/how-advisory-boards-can-help-your-startup
- RockHealth Video: www.youtube.com/watch?v=sTcpUCmUu4A
- FAST document:

  http://fi.co/contents/fast#
- John Finn jfinn@geiconsultants.com

GEI Consultants

392021 The I-Corps Program at NSF (and NIH)

Abstract:

There have been profound recent developments that have changed the way universities and the nation's funding agencies foster the creation of startups. Come learn about this new approach, how the NSF and NIH are implementing it in their new I-Corps program, and what this means to you as you seek to commercialize your innovation

AUTHOR INDEX

Borgese, Christina M. **14, 63**
Cabano, Stephen L. **6**
Colville, Alexander **32**
Dunn, Russell F. **62**
Figueroa, Nancy **29**
Finn, John **69**
Greer, Jessica **18**
Hipple, Jack **1, 64**
Larsen, Eldon R. **1, 15, 17**
McCullough, Jack **31**
McGreggor, Keith **79**
Nordin, Nadiah **30**
Privitera, Marc **16, 67**
Roscetti, Jennifer **5, 68**
Wang, Zixuan **19**
Watkins, Dan **61**
White, Sam **33**
Yuill, Barbara **66**